T0174169

Random Circulant Matrices

Random Circulant Matrices

By
Arup Bose
Koushik Saha

CRC Press
Taylor & Francis Group
Boca Raton London New York

CRC Press is an imprint of the
Taylor & Francis Group, an **informa** business
A CHAPMAN & HALL BOOK

CRC Press
Taylor & Francis Group
6000 Broken Sound Parkway NW, Suite 300
Boca Raton, FL 33487-2742

First issued in paperback 2020

© 2019 by Taylor & Francis Group, LLC
CRC Press is an imprint of Taylor & Francis Group, an Informa business

No claim to original U.S. Government works

Version Date: 20180913

ISBN 13: 978-0-367-73291-2 (pbk)
ISBN 13: 978-1-138-35109-7 (hbk)

This book contains information obtained from authentic and highly regarded sources. Reasonable efforts have been made to publish reliable data and information, but the author and publisher cannot assume responsibility for the validity of all materials or the consequences of their use. The authors and publishers have attempted to trace the copyright holders of all material reproduced in this publication and apologize to copyright holders if permission to publish in this form has not been obtained. If any copyright material has not been acknowledged please write and let us know so we may rectify in any future reprint.

Except as permitted under U.S. Copyright Law, no part of this book may be reprinted, reproduced, transmitted, or utilized in any form by any electronic, mechanical, or other means, now known or hereafter invented, including photocopying, microfilming, and recording, or in any information storage or retrieval system, without written permission from the publishers.

For permission to photocopy or use material electronically from this work, please access www.copyright.com (http://www.copyright.com/) or contact the Copyright Clearance Center, Inc. (CCC), 222 Rosewood Drive, Danvers, MA 01923, 978-750-8400. CCC is a not-for-profit organization that provides licenses and registration for a variety of users. For organizations that have been granted a photocopy license by the CCC, a separate system of payment has been arranged.

Trademark Notice: Product or corporate names may be trademarks or registered trademarks, and are used only for identification and explanation without intent to infringe.

Library of Congress Cataloging-in-Publication Data

Names: Bose, Arup, author. | Saha, Koushik, author.
Title: Random circulant matrices / Arup Bose and Koushik Saha.
Description: Boca Raton : CRC Press, Taylor & Francis Group, 2018.
Identifiers: LCCN 2018028758 | ISBN 9781138351097 (hardback)
Subjects: LCSH: Random matrices--Problems, exercises, etc. |
Matrices--Problems, exercises, etc. | Eigenvalues--Problems, exercises, etc.
Classification: LCC QA196.5 .B6725 2018 | DDC 512.9/434--dc23
LC record available at https://lccn.loc.gov/2018028758

Visit the Taylor & Francis Web site at
http://www.taylorandfrancis.com

and the CRC Press Web site at
http://www.crcpress.com

To the memory of my teacher J.C. Gupta

A.B.

To my parents and teachers

K.S.

Contents

Preface   xi

About the Authors   xiii

Introduction   xv

1 Circulants   1
   1.1 Circulant 1
   1.2 Symmetric circulant   2
   1.3 Reverse circulant   3
   1.4 k-circulant   3
   1.5 Exercises 8

2 Symmetric and reverse circulant   9
   2.1 Spectral distribution   9
   2.2 Moment method   10
        2.2.1 Scaling 12
        2.2.2 Input and link   12
        2.2.3 Trace formula and circuits   13
        2.2.4 Words and vertices   14
        2.2.5 (M1) and Riesz's condition   16
        2.2.6 (M4) condition   16
   2.3 Reverse circulant   17
   2.4 Symmetric circulant   20
   2.5 Related matrices 22
   2.6 Reduced moment 23
        2.6.1 A metric   23
        2.6.2 Minimal condition   24
   2.7 Exercises 25

3 LSD: normal approximation   27
   3.1 Method of normal approximation   27
   3.2 Circulant   28
   3.3 k-circulant   31
   3.4 Exercises 35

4   LSD: dependent input   **37**
   4.1   Spectral density   37
   4.2   Circulant   38
   4.3   Reverse circulant   45
   4.4   Symmetric circulant   47
   4.5 k-circulant   50
   4.6   Exercises   65

5   Spectral radius: light tail   **67**
   5.1   Circulant and reverse circulant   67
   5.2   Symmetric circulant   70
   5.3   Exercises   78

6   Spectral radius: k-circulant   **79**
   6.1   Tail of product   79
   6.2   Additional properties of the k-circulant   83
   6.3   Truncation and normal approximation   86
   6.4   Spectral radius of the k-circulant   88
        6.4.1 k-circulant for $sn = k^g + 1$ 97
   6.5   Exercises   98

7   Maximum of scaled eigenvalues: dependent input   **99**
   7.1   Dependent input with light tail   99
   7.2   Reverse circulant and circulant   100
   7.3   Symmetric circulant   104
   7.4 k-circulant   114
        7.4.1 k-circulant for $n = k^2 + 1$ 115
        7.4.2 k-circulant for $n = k^g + 1, g > 2$ 117
   7.5   Exercises   118

8   Poisson convergence   **119**
   8.1   Point process   119
   8.2   Reverse circulant   120
   8.3   Symmetric circulant   126
   8.4 k-circulant, $n = k^2 + 1$ 128
   8.5   Reverse circulant: dependent input   135
   8.6   Symmetric circulant: dependent input   137
   8.7 k-circulant, $n = k^2 + 1$: dependent input   137
   8.8   Exercises   138

9   Heavy-tailed input: LSD   **139**
   9.1   Stable distribution and input sequence   139
   9.2   Background   140
   9.3   Reverse circulant and symmetric circulant   144
   9.4 k-circulant: $n = k^g + 1$ 145
        9.4.1   Proof of Theorem 9.4.2   149

9.5 k-circulant: $n = k^g - 1$ 154
9.6 Tail of the LSD 156
9.7 Exercises 157

10 Heavy-tailed input: spectral radius **159**
10.1 Input sequence and scaling 159
10.2 Reverse circulant and circulant 160
10.3 Symmetric circulant 164
10.4 Heavy-tailed: dependent input 166
10.5 Exercises 171

11 Appendix **173**
11.1 Proof of Theorem 1.4.1 173
11.2 Standard notions and results 177
11.3 Three auxiliary results 182

Bibliography **185**

Index **189**

Preface

Circulant matrices have been around for a long time and have been extensively used in many scientific areas. The classic book *Circulant Matrices* by P. Davis, has a wealth of information on these matrices. New research on, and applications of, these matrices are continually appearing everywhere.

This book studies the properties of the eigenvalues for various types of circulant matrices—usual circulant, reverse circulant, and k-circulant when the dimension of the matrices grow, and the entries are random. The behavior of the spectral distribution, the spectral radius, and the appropriate point processes are developed systematically using the method of moments, and normal approximation results. This behavior varies according as the entries are independent, are from a linear process, and are light- or heavy-tailed.

The eigenvalues of these matrices can be calculated explicitly or can be described by a combinatorial formula. We take advantage of this formula to describe the asymptotic behavior of the eigenvalues.

The limiting spectral distribution (LSD) of these matrices are described using some functions of the Gaussian distribution. For those matrices which are symmetric, these limits are established using the moment method. For non-symmetric matrices we use the method of normal approximation.

With the help of sharp normal approximation results, it is shown that the spectral radius has Gumbel-type limit. With a natural labeling of the eigenvalues that uses the discrete Fourier frequencies, the point processes defined by the eigenvalues converge to Poisson point processes.

Some of the above results are generalized to the situation where the entries are not i.i.d. but are from a stationary linear process. The LSDs are mixtures of Gaussian, their functions, and the spectral density of the linear process. The limit distribution of the spectral radius can only be obtained after suitable scaling of the individual eigenvalues. We also demonstrate what happens when the entries are in the domain of attraction of a heavy-tailed distribution.

To keep the material focused and widely accessible, many topics have been omitted. Block circulants, period m-circulants, sparse circulants, and the role of the random circulant in the study of random Toeplitz matrices, are conspicuous by their absence.

Much of the material is based on collaborative research between Rajat Subhra Hazra, Arnab Sen, Joydip Mitra, and us. We joyfully recall the good times.

The work of AB was supported in part by the J.C. Bose Fellowship of the Govt. of India. The work of KS was supported in part by the Seed Grant of IIT Bombay.

It is always a pleasure to work with the Acquiring Editor John Kimmel, and the super-efficient go-to person for all our LaTeX needs, Mr. Sashi Kumar.

Arup Bose
Kolkata, India

Koushik Saha
Mumbai,India

About the Authors

Arup Bose earned his B.Stat., M.Stat. and Ph.D. degrees from the Indian Statistical Institute. He has been on its faculty at the Theoretical Statistics and Mathematics Unit, Kolkata, India, since 1991. He is a Fellow of the Institute of Mathematical Statistics, and of all three national science academies of India. He is a recipient of the S.S. Bhatnagar Prize and the C.R. Rao Award. He is the author of three books—*Patterned Random Matrices, Large Covariance and Autocovariance Matrices* (with Monika Bhattacharjee) and *U-Statistics, M_m-Estimators and Resampling* (with Snigdhansu Chatterjee).

Koushik Saha earned a B.Sc. in Mathematics from the Ramakrishna Mission Vidyamandira, Belur and an M.Sc. in Mathematics from the Indian Institute of Technology Bombay. He obtained his Ph.D. degree from the Indian Statistical Institute under the supervision of Arup Bose. His thesis on circulant matrices received high praise from the reviewers. He has been on the faculty of the Department of Mathematics, Indian Institute of Technology Bombay since 2014.

Introduction

Circulant matrices have been around for a long time and have been extensively used in many scientific areas. One may recall the classic book by Davis (1979) on these matrices with fixed entries. New research and applications on these matrices are continually appearing everywhere.

We focus on the usual *circulant* (C_n), and their close cousins such as the *symmetric circulant* (SC_n), the *reverse circulant* (RC_n) and the *k-circulant* matrices when the entries are random and the dimension of the matrix increases to infinity. Collectively, they will be labelled as "*circulant-type matrices*". We collect together the variety of results that are available on the behavior of the eigenvalues of these matrices.

An $n \times n$ *k-circulant* matrix is defined as

$$A_{k,n} = \begin{bmatrix} x_0 & x_1 & x_2 & \cdots & x_{n-2} & x_{n-1} \\ x_{n-k} & x_{n-k+1} & x_{n-k+2} & \cdots & x_{n-k-2} & x_{n-k-1} \\ x_{n-2k} & x_{n-2k+1} & x_{n-2k+2} & \cdots & x_{n-2k-2} & x_{n-2k-1} \\ & & & \vdots & & \\ x_k & x_{k+1} & x_{k+2} & \cdots & x_{k-2} & x_{k-1} \end{bmatrix}_{n \times n}.$$

The sequence $\{x_i\}$ is called the *input* sequence. The subscripts of the entries are to be read modulo n. For $1 \leq j < n - 1$, its $(j + 1)$-th row is obtained by giving its j-th row a right circular shift by $k \mod n$ positions. For the special cases of $k = 1$ and $k = n - 1$, they are known as the circulant matrix (C_n) and the reverse circulant matrix (RC_n) respectively. If symmetry is imposed on C_n, then it is called the symmetric circulant (SC_n).

There are many branches of science where these matrices play important roles. Davis (1979) is a rich source for results in and applications of circulant matrices which have non-random entries. See also Pollock (2002). Here are a few scattered examples of the uses of circulant matrices. The periodogram of a sequence $\{a_l\}_{l \geq 0}$ is defined as $n^{-1}|\sum_{l=0}^{n-1} a_l e^{2\pi i j/n}|^2$, $-\lfloor \frac{n-1}{2} \rfloor \leq j \leq \lfloor \frac{n-1}{2} \rfloor$. Properties of the periodogram are fundamental in the spectral analysis of time series. See for instance Brockwell and Davis (2006), and Fan and Yao (2003). It is easy to see that the periodogram is a straightforward function of the eigenvalues of a suitable circulant matrix.

The k-circulant matrices and their block versions arise in multi-level supersaturated design of experiment (see Georgiou and Koukouvinos (2006)), spectra of De Bruijn graphs (see Strok (1992)) and $(0, 1)$-matrix solutions to $A^m = J_n$ (see Wu et al. (2002)).

Non-random Toeplitz matrices and the corresponding Toeplitz operators are well-studied objects in mathematics. Toeplitz matrices appear in several places—as the covariance matrix of stationary processes, in shift-invariant linear filtering and in many aspects of combinatorics, time series and harmonic analysis. The matrix C_n plays a crucial role in the study of large dimensional Toeplitz matrices with non-random input. See, for example, Grenander and Szegő (1984) and Gray (2006).

As mentioned earlier, our aim is to study the eigenvalue properties of the circulant-type matrices. If $\lambda_1, \ldots, \lambda_n$ are the eigenvalues of any $n \times n$ matrix A_n, then its *empirical spectral distribution function* (ESD) is given by

$$F_{A_n}(x, y) = \frac{1}{n} \sum_{i=1}^{n} \mathbb{I}\{\mathcal{R}(\lambda_i) \leq x, \ \mathcal{I}(\lambda_i) \leq y\}.$$

If the entries of A_n are random and $n \to \infty$, then these are called large dimensional random matrices (LDRM). The behavior of the eigenvalues of LDRMs, from many different angles, has attracted considerable interest in physics, mathematics, statistics, wireless communication and other branches of sciences.

The *limiting spectral distribution* (LSD) of the sequence $\{A_n\}$ is defined as the weak limit of the sequence $\{F_{A_n}\}$, if it exists. LSDs for different random matrix models have been established over the years. In particular, the LSD of a sequence of random symmetric Toeplitz matrices exists under appropriate conditions. This LSD is absolutely continuous with respect to the Lebesgue measure and quite interestingly, the embedding of the Toeplitz matrix into a circulant matrix has been crucially used to establish this. Another aspect that is studied for LDRMs is the limiting behavior near the "edge"—of the extreme eigenvalues and the spectral radius. This behavior of the extreme eigenvalues and related quantities is very non-trivial for most random matrices. Similarly, the spacings between the eigenvalues are also difficult objects to analyze for most random matrix models. We shall study all of these aspects for the circulant-type matrices.

In Chapter 1, we describe the structure of the different circulant-type matrices and their eigenvalues. A formula solution for the eigenvalues of the general k-circulant is presented in terms of the input sequence. Its proof is technical and is relegated to Appendix. Though this solution is in a reasonable closed form, considerable work is needed to turn it into results on the LSD, the spectral radius and the spacings of eigenvalues. These results are developed in a systematic and precise way, using the moment method and the method of normal approximation.

In Chapter 2 first a general approach, based on the moment method, is discussed in the context of LSD of symmetric random matrices. This is used to establish the LSDs of the two circulants that are symmetric, namely RC_n and SC_n, when the input is i.i.d.

In Chapter 3, we introduce the basic normal approximation technique and use it to obtain the LSDs of the non-symmetric circulants, namely C_n and $A_{k,n}$ (for certain combinations of k and n) when the input is independent.

In Chapter 4 the above results are extended to inputs which are stationary two-sided moving average processes of infinite order, that is,

$$x_n = \sum_{i=-\infty}^{\infty} a_i \varepsilon_{n-i}, \quad \text{where} \quad a_n \in \mathbb{R} \quad \text{and} \quad \sum_{n \in \mathbb{Z}} |a_n| < \infty. \tag{0.1}$$

The limits are suitable mixtures of normal, symmetric square root of the chi-square, and other distributions, with the spectral density of the process involved in the mixtures. For instance, under some conditions on $\{x_n\}$, the ESD of $\frac{1}{\sqrt{n}} RC_n$ converges weakly to the distribution F_R, where

$$F_R(x) = \begin{cases} 1 - \displaystyle\int_0^{1/2} e^{-\frac{x^2}{2\pi f(2\pi t)}} \, dt & \text{if } x > 0 \\[3mm] \displaystyle\int_0^{1/2} e^{-\frac{x^2}{2\pi f(2\pi t)}} \, dt & \text{if } x \le 0, \end{cases}$$

and f is the spectral density function of $\{x_n\}$.

In Chapter 5, sharper normal approximation results are introduced as the main tools. These are used to derive results on the spectral radius of scaled C_n, RC_n and SC_n matrices when the input sequence is i.i.d. with finite moments of suitable order. The almost sure and the distributional convergence of the spectral radius of RC_n and C_n are established. For instance, suppose that $\{x_i\}$ is i.i.d. with mean μ and $E|x_i|^{2+\delta} < \infty$ for some $\delta > 0$, and consider the RC_n with this input. Then

$$\frac{\mathrm{sp}(RC_n) - |\mu|n}{\sqrt{n}} \xrightarrow{\mathcal{D}} N(0,1) \text{ if } \mu \ne 0,$$

and

$$\frac{\mathrm{sp}(\frac{1}{\sqrt{n}} RC_n) - d_q}{c_q} \xrightarrow{\mathcal{D}} \Lambda \text{ if } \mu = 0,$$

where

$$q = q(n) = \lfloor \frac{n-1}{2} \rfloor, \quad d_q = \sqrt{\ln q}, \quad c_q = \frac{1}{2\sqrt{\ln q}},$$

and Λ is the standard Gumbel distribution. The joint behavior of the minimum and the maximum eigenvalue of SC_n and inter-alia the distributional convergence of the spectral radius are also established in this chapter.

In Chapter 6, we first identify the tail behavior of the product of i.i.d. exponential random variables. Suppose

$$H_m(x) = P[E_1 E_2 \cdots E_m > x],$$

where $\{E_i\}$ are i.i.d. standard exponential random variables. First we show that for any m, $1 - H_m(\cdot)$ lies in the maximum domain of attraction of the Gumbel distribution. Using this result and significantly extending the ideas of the previous chapter, the limit distribution of the spectral radius of the k-circulant when $n = k^g + 1$, $g \geq 2$, and when $sn = k^g + 1$ with some suitable condition on s is derived.

Now suppose that the input sequence is a stationary two-sided moving average process of infinite order. For such an input sequence, the eigenvalues are scaled by the spectral density at appropriate ordinates. The limit behavior of the maximum of the modulus, say M, of these scaled eigenvalues is derived in Chapter 7. For instance, suppose $\{x_n\}$ is the two-sided moving average process as in (0.1) and

$$M(\frac{1}{\sqrt{n}}RC_n, f) = \max_{1 \leq k \leq n} \frac{|\lambda_k|}{\sqrt{2\pi f(\omega_k)}},$$

where λ_k are the eigenvalues of $\frac{1}{\sqrt{n}}RC_n$ and f is the spectral density of $\{x_n\}$. Then under some assumptions on $\{x_n\}$,

$$\frac{M(\frac{1}{\sqrt{n}}RC_n, f) - d_q}{c_q} \xrightarrow{\mathcal{D}} \Lambda,$$

where $q = q(n) = \lfloor \frac{n-1}{2} \rfloor$, $d_q = \sqrt{\ln q}$ and $c_q = \frac{1}{2\sqrt{\ln q}}$.

In Chapter 8, the convergence of the point process based on the eigenvalues of RC_n, SC_n and k-circulant with i.i.d. entries is discussed. For example, consider RC_n. Since most of the eigenvalues of RC_n appear in pairs with opposite signs, we consider the following point process based on half of them:

$$\eta_n(\cdot) = \sum_{j=0}^{q} \epsilon_{(\omega_j, \frac{\lambda_j - b_q}{a_q})}(\cdot),$$

where $\{\lambda_j\}$ are the eigenvalues (labelled in a specified way) of $\frac{1}{\sqrt{n}}RC_n$, $\{\omega_j = \frac{2\pi j}{n}\}$ are the Fourier frequencies, a_q, b_q are appropriate scaling and centering constants, and $q = \lfloor \frac{n}{2} \rfloor$. Then under some conditions on the input, $\eta_n \xrightarrow{\mathcal{D}} \eta$, where η is a Poisson process on $[0, \pi] \times (-\infty, \infty]$ with intensity function $\lambda(t, x) = \pi^{-1}e^{-x}$. Joint convergence of the upper k-ordered eigenvalues and their spacings follow from this result.

In Chapter 9 the LSD of a class of circulant-type random matrices with heavy-tailed input is considered. Unlike the light-tailed case where the limit is non-random, here the limit is a random probability distribution. An explicit representation of the limit is provided.

The distributional convergence of the spectral radius of the scaled eigenvalues of C_n, RC_n and SC_n when the input sequence is i.i.d. with appropriate heavy tail, is considered in Chapter 10. For instance, consider RC_n with an input sequence $\{Z_t, \ t \in \mathbb{Z}\}$ which is i.i.d. with a common distribution

F in the *domain of attraction* of an α-*stable* distribution with $0 < \alpha < 1$. Then $\mathbf{sp}(\frac{1}{b_n} RC_n) \xrightarrow{\mathcal{D}} Y_\alpha$, where Y_α has stable distribution with index α and $b_n \approx n^{1/\alpha} L_0(n)$ for some slowly varying function L_0. In this heavy tail situation the limit distribution is not Gumbel anymore. We also consider the maximum of the modulus of scaled eigenvalues for RC_n, C_n and SC_n matrices when the input is a moving average with heavy-tailed noise sequence, and establish its weak limit.

In Appendix, we have established in detail an eigenvalue formula solution for the k-circulant. We have also briefly defined and explained some of the background material in probability theory that is needed in the main text.

1

Circulants

There are several reasons to study the circulant matrix and its variants. For example, the large dimensional non-random Toeplitz matrix and the corresponding Toeplitz operator are well-studied objects in mathematics. The non-random circulant matrix plays a crucial role in these studies. See, for example, Grenander and Szegő (1984) and Gray (2006). The eigenvalues of the circulant matrix also arise crucially in time series analysis. For instance, the periodogram of a sequence $\{a_l\}_{l \geq 0}$ is defined as $n^{-1}|\sum_{l=0}^{n-1} a_l e^{2\pi i j l/n}|^2$, $-\lfloor \frac{n-1}{2} \rfloor \leq j \leq \lfloor \frac{n-1}{2} \rfloor$ and it is a straightforward function of the eigenvalues of a suitable circulant matrix. The properties of the periodogram are fundamental in the spectral analysis of time series. See for instance Fan and Yao (2003). The maximum of the periodogram, in particular, has been studied in Davis and Mikosch (1999). The k-circulant matrices and their block versions arise in areas such as multi-level supersaturated design of experiment (Georgiou and Koukouvinos (2006)), spectra of De Bruijn graphs (Strok (1992)) and $(0, 1)$-matrix solutions to $A^m = J_n$ (Wu et al. (2002)). See also Davis (1979) and Pollock (2002).

In this chapter we will define all the variants of the circulant matrix that we shall discuss in this book. The nice thing about these circulants is that formula solutions are known for their eigenvalues. These formulae are simple for some of the matrices and much harder to describe for some others. We provide detailed derivation of these formulae. Except in the next chapter, the developments in all the subsequent chapters rely heavily on these formulae.

1.1 Circulant

The usual *circulant matrix* is defined as

$$C_n = \begin{bmatrix} x_0 & x_1 & x_2 & \cdots & x_{n-2} & x_{n-1} \\ x_{n-1} & x_0 & x_1 & \cdots & x_{n-3} & x_{n-2} \\ x_{n-2} & x_{n-1} & x_0 & \cdots & x_{n-4} & x_{n-3} \\ & & & \vdots & & \\ x_1 & x_2 & x_3 & \cdots & x_{n-1} & x_0 \end{bmatrix}_{n \times n}.$$

For $1 \leq j < n - 1$, its $(j + 1)$-th row is obtained by giving its j-th row a right circular shift by one position and the (i, j)-th element of the matrix is $x_{(j-i+n) \bmod n}$. It is easy to see that irrespective of the entries, the eigenvectors of the circulant matrix are $(1, \omega, \ldots, \omega^{n-1})^T$ where ω is any n-th root of unity. Its eigenvalues are also known explicitly. Define

$$\omega_k = \frac{2\pi k}{n} \quad \text{for } k \geq 0 \text{ and } i = \sqrt{-1}. \tag{1.1}$$

Then the eigenvalues of C_n are given by (see, for example, Brockwell and Davis (2006)):

$$\lambda_k = \sum_{j=0}^{n-1} x_j \exp(i\omega_k j)$$

$$= \sum_{j=0}^{n-1} x_j \cos(\omega_k j) + i \sum_{j=0}^{n-1} x_j \sin(\omega_k j), \quad 0 \leq k \leq n - 1. \tag{1.2}$$

1.2 Symmetric circulant

This is really the symmetric version of C_n and is defined as

$$SC_n = \begin{bmatrix} x_0 & x_1 & x_2 & \cdots & x_2 & x_1 \\ x_1 & x_0 & x_1 & \cdots & x_3 & x_2 \\ x_2 & x_1 & x_0 & \cdots & x_2 & x_3 \\ & & & \vdots & & \\ x_1 & x_2 & x_3 & \cdots & x_1 & x_0 \end{bmatrix}_{n \times n}.$$

The first row $(x_0\ x_1\ x_2 \cdots x_2\ x_1)$ is a palindrome and the $(j + 1)$-th row is obtained by giving its j-th row a right circular shift by one position. Its (i, j)-th element is given by $x_{n/2 - |n/2 - |i-j||}$. Specializing the general formula for the eigenvalues of C_n to SC_n, its eigenvalues are given by:

(a) for n odd:

$$\lambda_0 = x_0 + 2 \sum_{j=1}^{\lfloor \frac{n}{2} \rfloor} x_j,$$

$$\lambda_k = x_0 + 2 \sum_{j=1}^{\lfloor \frac{n}{2} \rfloor} x_j \cos(\omega_k j), \quad 1 \leq k \leq \lfloor \frac{n}{2} \rfloor,$$

with $\lambda_{n-k} = \lambda_k$ for $1 \leq k \leq \lfloor \frac{n}{2} \rfloor$. Here $\lfloor x \rfloor$ denotes the greatest integer that is less than or equal to x.

(b) for n even:

$$\lambda_0 = x_0 + 2 \sum_{j=1}^{\frac{n}{2}-1} x_j + x_{n/2},$$

$$\lambda_k = x_0 + 2 \sum_{j=1}^{\frac{n}{2}-1} x_j \cos(\omega_k j) + (-1)^k x_{n/2}, \quad 1 \le k \le \frac{n}{2},$$

with $\lambda_{n-k} = \lambda_k$ for $1 \le k \le \frac{n}{2}$.

1.3 Reverse circulant

The so-called *reverse circulant* matrix is given by

$$RC_n = \begin{bmatrix} x_0 & x_1 & x_2 & \cdots & x_{n-2} & x_{n-1} \\ x_1 & x_2 & x_3 & \cdots & x_{n-1} & x_0 \\ x_2 & x_3 & x_4 & \cdots & x_0 & x_1 \\ & & & \vdots & & \\ x_{n-1} & x_0 & x_1 & \cdots & x_{n-3} & x_{n-2} \end{bmatrix}_{n \times n}.$$

For $1 \le j < n - 1$, its $(j + 1)$-th row is obtained by giving its j-th row a *left* circular shift by one position. This is a symmetric matrix and its (i, j)-th element is given by $x_{(i+j-2) \bmod n}$.

The eigenvalues of RC_n are given by the following formula which is a special case of the general k-circulant matrix treated in the next section.

$$\lambda_0 = \sum_{j=0}^{n-1} x_j,$$

$$\lambda_{\frac{n}{2}} = \sum_{j=0}^{n-1} (-1)^j x_j, \quad \text{if } n \text{ is even, and}$$

$$\lambda_k = -\lambda_{n-k} = |\sum_{j=0}^{n-1} x_j \exp(i\omega_k j)|, \quad 1 \le k \le \lfloor \frac{n-1}{2} \rfloor. \tag{1.3}$$

1.4 k-circulant

This is a k-shift generalization of C_n. For positive integers k and n, the $n \times n$ k-circulant matrix is defined as

$$A_{k,n} = \begin{bmatrix} x_0 & x_1 & x_2 & \cdots & x_{n-2} & x_{n-1} \\ x_{n-k} & x_{n-k+1} & x_{n-k+2} & \cdots & x_{n-k-2} & x_{n-k-1} \\ x_{n-2k} & x_{n-2k+1} & x_{n-2k+2} & \cdots & x_{n-2k-2} & x_{n-2k-1} \\ & & \vdots & & & \\ x_k & x_{k+1} & x_{k+2} & \cdots & x_{k-2} & x_{k-1} \end{bmatrix}_{n \times n}.$$

We emphasize that all subscripts appearing above are calculated modulo n. For $1 \le j < n-1$, its $(j+1)$-th row is obtained by giving its j-th row a right circular shift by k positions (equivalently, $k \bmod n$ positions).

It is clear that C_n and RC_n are special cases of the k-circulant when we let $k = 1$ and $k = n-1$, respectively.

We now give a brief description of the eigenvalues. For any positive integers k, n, let $p_1 < p_2 < \cdots < p_c$ be all their common prime factors so that

$$n = n' \prod_{q=1}^{c} p_q^{\beta_q} \quad \text{and} \quad k = k' \prod_{q=1}^{c} p_q^{\alpha_q}. \tag{1.4}$$

Here α_q, $\beta_q \ge 1$ and n', k', p_q are pairwise relatively prime. For any positive integer m, let

$$\mathbb{Z}_m = \{0, 1, 2, \ldots, m-1\}.$$

For fixed k and n, define the sets

$$S(x) = \{xk^b \ (\bmod \ n') : b \text{ is an integer and } b \ge 0\}, \tag{1.5}$$

where $x \in \mathbb{Z}_{n'}$ and n' is as in (1.4).

We will use the following notation throughout the book. For any set A,

$$\#A = \text{cardinality of } A.$$

Let

$$g_x = \#S(x).$$

We call g_x the *order* of x. Note that $g_0 = 1$. We observe the following about the sets $S(x)$ and the numbers g_x.

(i) $S(x) = \{xk^b \ (\bmod \ n') : 0 \le b < g_x\}$.

(ii) An alternative description of g_x, which will be used later, is the following. For $x \in \mathbb{Z}_{n'}$, let

$$\mathcal{O}_x = \{b > 0 \ : b \text{ is an integer and } xk^b = x \ (\bmod \ n')\}. \tag{1.6}$$

Then $g_x = \min \mathcal{O}_x$, that is g_x is the smallest positive integer b such that $xk^b = x \ (\bmod \ n')$.

(iii) For $x \ne u$, either $S(x) = S(u)$ or $S(x) \cap S(u) = \emptyset$. As a consequence, the distinct sets from the collection $\{S(x) : x \in \mathbb{Z}_{n'}\}$ form a partition of $\mathbb{Z}_{n'}$.

We call the distinct sets from $\{S(x) : x \in \mathbb{Z}_{n'}\}$ the *eigenvalue partition* of $\mathbb{Z}_{n'}$ and we will denote the partitioning sets and their sizes by

$$\{\mathcal{P}_0 = \{0\}, \mathcal{P}_1, \ldots, \mathcal{P}_{l-1}\}, \quad \text{and} \quad n_j = \#\mathcal{P}_j, \ 0 \le j < l. \tag{1.7}$$

Define

$$y_j := \prod_{t \in \mathcal{P}_j} \lambda_{ty}, \quad j = 0, 1, \ldots, l - 1, \tag{1.8}$$

where $y = n/n'$ and λ_{ty} is as defined in (1.2).

We now provide the solution for the eigenvalues of $A_{k,n}$. In Appendix, we have reproduced its proof from Bose et al. (2012b).

Theorem 1.4.1 (Zhou (1996)). The characteristic polynomial of $A_{k,n}$ is given by

$$\chi(A_{k,n})(\lambda) = \lambda^{n-n'} \prod_{j=0}^{\ell-1} (\lambda^{n_j} - y_j), \tag{1.9}$$

where y_j is as defined in (1.8).

Now we note some useful properties of the eigenvalue partition $\{\mathcal{P}_j, j = 0, 1, \ldots, l-1\}$ in the following lemmata. The readers may ignore these properties for the time being but they will be used in Chapters 4, 6 and 8. A more detailed analysis of the eigenvalues, useful in deriving the limiting distribution of the spectral radius for a specific class of k-circulant matrices, has been developed in Section 6.2.

Lemma 1.4.2. (a) Let $x, y \in \mathbb{Z}_{n'}$. If $n' - t_0 \in S(y)$ for some $t_0 \in S(x)$, then for every $t \in S(x)$, we have $n' - t \in S(y)$.

(b) Fix $x \in \mathbb{Z}_{n'}$. Then g_x divides b for every $b \in \mathcal{O}_x$. Furthermore, g_x divides g_1 for each $x \in \mathbb{Z}_{n'}$.

(c) Suppose g divides g_1. Set $m := \gcd(k^g - 1, n')$. Let $X(g)$ and $Y(g)$ be defined as

$$X(g) := \Big\{ x : x \in \mathbb{Z}_{n'} \text{ and } x \text{ has order } g \Big\}, \ Y(g) := \Big\{ bn'/m \ : \ 0 \le b < m \Big\}.$$

Then

$$X(g) \subseteq Y(g), \quad \#Y(g) = m, \quad \text{and} \quad \bigcup_{h:h|g} X(h) = Y(g).$$

Proof. (a) Since $t \in S(x) = S(t_0)$, we can write $t = t_0 k^b \pmod{n'}$ for some $b \ge 0$. Therefore $n' - t = (n' - t_0)k^b \pmod{n'} \in S(n' - t_0) = S(y)$.

(b) Fix $b \in \mathcal{O}_x$. Since g_x is the smallest element of $\mathcal{O}_x$, it follows that $g_x \le b$. Suppose, if possible, $b = qg_x + r$ where $0 < r < g_x$. By the fact that $xk^{g_x} = x \pmod{n'}$, it then follows that

$$x = xk^b \pmod{n'} = xk^{qg_x + r} \pmod{n'} = xk^r \pmod{n'}.$$

This implies that $r \in \mathcal{O}_x$ and $r < g_x$, which is a contradiction to the fact that g_x is the smallest element in $\mathcal{O}_x$. Hence, we must have $r = 0$ proving that g_x divides b.

Note that $k^{g_1} = 1 \pmod{n'}$, implying that $xk^{g_1} = x \pmod{n'}$. Therefore $g_1 \in \mathcal{O}_x$ proving the assertion.

(c) Clearly, $\#Y(g) = m$. Fix $x \in X(h)$ where h divides g. Then, $xk^g = x(k^h)^{g/h} = x \pmod{n'}$, since g/h is a positive integer. Therefore n' divides $x(k^g - 1)$. So, n'/m divides $x(k^g - 1)/m$. But n'/m is relatively prime to $(k^g - 1)/m$ and hence n'/m divides x. So, $x = bn'/m$ for some integer $b \geq 0$. Since $0 \leq x < n'$, we have $0 \leq b < m$, and $x \in Y(g)$, proving $\bigcup_{h:h|g} X(h) \subseteq Y(g)$ and in particular, $X(g) \subseteq Y(g)$.

On the other hand, take $0 \leq b < g$. Then $(bn'/m) k^g = (bn'/m) \bmod n'$. Hence $g \in \mathcal{O}_{bn'/m}$, and which implies, by part (b) of the lemma, that $g_{cn'/m}$ divides g. Therefore $Y(g) \subseteq \bigcup_{h:h|g} X(h)$, and that completes the proof. □

Lemma 1.4.3. Let $g_1 = q_1^{\gamma_1} q_2^{\gamma_2} \cdots q_m^{\gamma_m}$ where $q_1 < q_2 < \cdots < q_m$ are primes. Define for $1 \leq j \leq m$,

$$L_j := \{q_{i_1} q_{i_2} \cdots q_{i_j} : 1 \leq i_1 < \cdots < i_j \leq m\}$$

and

$$G_j = \sum_{l_j \in L_j} \#Y(g_1/\ell_j) = \sum_{l_j \in L_j} \gcd\left(k^{g_1/\ell_j} - 1, n'\right).$$

Then we have

(a) $\# \{x \in \mathbb{Z}_{n'} : g_x < g_1\} = G_1 - G_2 + G_3 - G_4 + \cdots$.

(b) $G_1 - G_2 + G_3 - G_4 + \cdots \leq G_1$.

Proof. Fix $x \in \mathbb{Z}_{n'}$. By Lemma 1.4.2(b), g_x divides g_1 and hence we can write $g_x = q_1^{\eta_1} \cdots q_m^{\eta_m}$ where, $0 \leq \eta_b \leq \gamma_b$ for $1 \leq b \leq m$. Since $g_x < g_1$, there is at least one b so that $\eta_b < \gamma_b$. Suppose that exactly h-many η's are equal to the corresponding γ's where $0 \leq h < m$. To keep notation simple, we will assume that, $\eta_b = \gamma_b$, $1 \leq b \leq h$ and $\eta_b < \gamma_b$, $h + 1 \leq b \leq m$.

(a) Then $x \in Y(g_1/q_b)$ for $h+1 \leq b \leq m$ and $x \notin Y(g_1/q_b)$ for $1 \leq b \leq h$. So, x is counted $(m - h)$ times in G_1. Similarly, x is counted $\binom{m-h}{2}$ times in G_2, $\binom{m-h}{3}$ times in G_3, and so on. Hence the total number of times x is counted in $(G_1 - G_2 + G_3 - \cdots)$ is

$$\binom{m-h}{1} - \binom{m-h}{2} + \binom{m-h}{3} - \cdots = 1.$$

(b) Note that $m - h \geq 1$. Further, each element in the set $\{x \in \mathbb{Z}_{n'} : g_x < g_1\}$ is counted once in $G_1 - G_2 + G_3 - \cdots$ and $(m - h)$ times in G_1. The result follows immediately. □

In Lemma 1.4.2(b), we observed that $g_x \leq g_1$ for all $x \in \mathbb{Z}_{n'}$. We will now consider the elements in $\mathbb{Z}_{n'}$ whose orders are strictly less than g_1. We define

$$v_{k,n'} = \#\{x \in \mathbb{Z}_{n'} : g_x < g_1\}. \tag{1.10}$$

The following lemma establishes upper bounds on $v_{k,n'}$ and will be crucially used in Chapters 4 and 6.

Lemma 1.4.4. (a) If $g_1 = 2$, then $v_{k,n'} = \gcd(k-1, n')$.

(b) If $g_1 \geq 4$ is even and $k^{g_1/2} = -1 \pmod{n}$, then

$$v_{k,n'} \leq 1 + \sum_{b:\, b|g_1, b\geq 3} \gcd(k^{g_1/b} - 1, n').$$

(c) If $g_1 \geq 2$ and q_1 is the smallest prime divisor of g_1, then $v_{k,n'} < 2k^{g_1/q_1}$.

Proof. (a) This is immediate from Lemma 1.4.3(a) which asserts that $v_{k,n'} = \#Y(1) = \gcd(k-1, n')$.

(b) Fix $x \in \mathbb{Z}_{n'}$ with $g_x < g_1$. Since g_x divides g_1 and $g_x < g_1$, g_x must be of the form g_1/b for some integer $b \geq 2$ provided g_1/b is an integer.

If $b = 2$, then $xk^{g_1/2} = xk^{g_x} = x \pmod{n'}$. But $k^{g_1/2} = -1 \pmod{n'}$ and so, $xk^{g_1/2} = -x \pmod{n'}$. Therefore, $2x = 0 \pmod{n'}$ and x can be either 0 or $n'/2$, provided $n'/2$ is an integer. But $g_0 = 1 < 2 \leq g_1/2$ so x cannot be 0. So, there is at most one element in the set $X(g_1/2)$.

Thus, we have

$$\#\{x \in \mathbb{Z}_{n'} : g_x < g_1\} = \#X(g_1/2) + \sum_{b|g_1,\, b\geq 3} \#\{x \in \mathbb{Z}_{n'} : g_x = g_1/b\}$$

$$= \#X(g_1/2) + \sum_{b|g_1,\, b\geq 3} \#X(g_1/b)$$

$$\leq 1 + \sum_{b|g_1,\, b\geq 3} \#Y(g_1/b) \quad [\text{by Lemma 1.4.2(c)}]$$

$$= 1 + \sum_{b|g_1,\, b\geq 3} \gcd(k^{g_1/b} - 1, n') \quad [\text{by Lemma 1.4.2(c)}].$$

(c) As in Lemma 1.4.3, let $g_1 = q_1^{\gamma_1} q_2^{\gamma_2} \cdots q_m^{\gamma_m}$ where $q_1 < q_2 < \cdots < q_m$ are primes. Then by Lemma 1.4.3,

$$v_{k,n'} = G_1 - G_2 + G_3 - G_4 + \cdots \leq G_1 = \sum_{b=1}^{m} \gcd(k^{g_1/q_b} - 1, n')$$

$$< \sum_{b=1}^{m} k^{g_1/q_b}$$

$$\leq 2k^{g_1/q_1},$$

where the last inequality follows from the observation

$$\sum_{b=1}^{m} k^{g_1/q_b} \leq k^{g_1/q_1} \sum_{b=1}^{m} k^{-g_1(q_b-q_1)/q_1 q_b}$$

$$\leq k^{g_1/q_1} \sum_{b=1}^{m} k^{-(q_b-q_1)}$$

$$\leq k^{g_1/q_1} \sum_{b=1}^{m} k^{-(b-1)}$$

$$\leq 2k^{g_1/q_1}.$$

$\square$

1.5   Exercises

1. Check that $(1, \omega, \ldots, \omega^{n-1})^T$ where ω is any n-th root of unity are indeed the eigenvectors of C_n and (1.2) are indeed its eigenvalues.

2. Check that when we specialize to the cases $k = 1$ and $k = n-1$, the k-circulant reduces to C_n and RC_n, respectively. Also check that the eigenvalue formula for the k-circulant reduces respectively to the eigenvalue formula given for C_n in (1.2), and for RC_n in (1.3).

3. Show that circulant matrices of the same order commute.

4. Show that the k-circulant matrix is symmetric for all inputs if and only if $k = n - 1$.

5. Show that $S(x) = \{xk^b \pmod{n'} : 0 \leq b < \#S(x)\}$, where $S(x)$ is as defined in (1.5).

6. Prove that $g_x = \min \mathcal{O}_x$ where $\mathcal{O}_x$ is as defined in (1.6).

7. For $x \neq u$, show that either $S(x) = S(u)$ or $S(x) \cap S(u) = \emptyset$.

2

Symmetric and reverse circulant

We begin the chapter with the notions of empirical and limiting spectral distribution of large random matrices. One of the most useful and common methods to establish the limiting spectral distribution of a sequence of real symmetric matrices is the moment method. There are two circulant-type matrices which are symmetric, namely SC_n and RC_n. We establish the limiting spectral distribution of these two matrices by the moment method. The limits are universal; that is, they do not depend on the underlying distribution of the entries. The non-symmetric circulants will be tackled in the later chapters.

2.1   Spectral distribution

Entries of all our matrices are real in general. Suppose A_n is any $n \times n$ matrix with eigenvalues $\lambda_1, \lambda_2, \ldots, \lambda_n$ (written in some order).

Definition 2.1.1. The probability distribution which puts mass $\frac{1}{n}$ at each λ_i is known as the *empirical spectral measure* of A_n. The corresponding distribution function is called its *empirical spectral distribution (ESD)*.

Clearly, there can be complex eigenvalues and the ESD of A_n equals

$$F_{A_n}(x, y) = \frac{1}{n} \sum_{i=1}^{n} \mathbb{I}\{\mathcal{R}(\lambda_i) \leq x, \ \mathcal{I}(\lambda_i) \leq y\},$$

where $\mathcal{R}(z)$ and $\mathcal{I}(z)$ denote the real and imaginary parts of z. If the eigenvalues are all real then this ESD is defined on $\mathbb{R}$ by

$$F_{A_n}(x) = \frac{1}{n} \sum_{i=1}^{n} \mathbb{I}\{\lambda_i \leq x\}.$$

If A_n has random entries, then the ESD is a *random probability measure*.

The *expected spectral distribution function* of A_n is defined as $\mathrm{E}(F_n(\cdot))$. This expectation always exists and is a distribution function. The corresponding probability distribution is known as the *expected spectral measure*.

Definition 2.1.2. Let $\{A_n\}$ be a sequence of deterministic square matrices with ESD $\{F_{A_n}\}$. The *limiting spectral distribution* (LSD) of the sequence is defined as the weak limit of the sequence $\{F_{A_n}\}$, if it exists.

If the entries of $\{A_n\}$ are random, the above weak limit must be understood in some probabilistic sense. One might think of the following two options for the weak convergence of a sequence of random probability measures.

When the entries of $\{A_n\}$ are random variables defined on some probability space $(\Omega, \mathcal{F}, \mathrm{P})$, $\{F_{A_n}(\cdot)\}$ (on $\mathbb{R}$ or on $\mathbb{R}^2$ as the case may be) are random functions (of $\omega \in \Omega$) but we suppress this dependence. Let F be a non-random distribution function, on $\mathbb{R}$ or on $\mathbb{R}^2$ as the case may be. Let

$$C_F = \{t : t \text{ is a continuity point of } F\}.$$

Definition 2.1.3. (i) Say that the ESD of A_n converges to F *almost surely* if for almost every $\omega \in \Omega$ and for all $t \in C_F$, $F_{A_n}(t) \to F(t)$ as $n \to \infty$.

(ii) Say that the ESD of A_n converges to F *in probability* if, for all $\epsilon > 0$ and $t \in C_F$, $\mathrm{P}(|F_{A_n}(t) - F(t)| > \epsilon) \to 0$ as $n \to \infty$.

Since (ii) is equivalent to saying that for all $t \in C_F$,

$$\int_\Omega \left[F_{A_n}(t) - F(t) \right]^2 d\,\mathrm{P}(\omega) \to 0 \text{ as } n \to \infty,$$

we also say that F_{A_n} converges to F in L_2.

It is also easy to see that $(i) \Rightarrow (ii)$. Several methods to establish the LSD of a sequence of random matrices are known in the literature. Out of these, the moment method is the most suitable to deal with real symmetric matrices. We now discuss this method. Then we shall use it to establish results on the LSD of SC_n and RC_n.

2.2 Moment method

For any real-valued random variable X or its distribution F, $\beta_h(F)$ and $\beta_h(X)$ respectively will denote its h-th moment. The following result is well-known in the literature of weak convergence of probability measures. Its proof is easy and we leave it as an exercise.

Lemma 2.2.1. Let $\{Y_n\}$ be a sequence of real-valued random variables with distributions $\{F_n\}$. Suppose that $\beta_h(Y_n) \to \beta_h$ (say) for every positive integer h. If there is a unique distribution F whose moments are $\{\beta_h\}$, then Y_n (or equivalently F_n) converges to F in distribution.

Observe that the above lemma requires a uniqueness condition. The following result of Riesz (1923) offers a sufficient condition for this uniqueness and will be enough for our purposes. A proof can be found in Bose (2018) and in Bai and Silverstein (2010).

Lemma 2.2.2. Let $\{\beta_k\}$ be the sequence of moments of the distribution function F. Then F is the unique distribution with these moments if

$$\liminf_{k\to\infty} \frac{1}{k} \beta_{2k}^{\frac{1}{2k}} < \infty \text{ (Riesz's condition)}. \tag{2.1}$$

The moments of any normal random variable satisfy Riesz's condition (2.1). This fact will be useful to us and we record it as a lemma. It can be proven using Stirling's approximation for factorials and is left as an exercise.

Lemma 2.2.3. $\{\beta_k\}$ satisfies condition (2.1) if for some $0 < \Delta < \infty$,

$$\beta_{2k} \leq \frac{(2k)!}{k!2^k} \Delta^k, \quad k = 0, 1, \dots.$$

Now consider the ESD F_{A_n}. Its h-th moment has the following nice form:

$$\beta_h(F_{A_n}) = \frac{1}{n} \sum_{i=1}^{n} \lambda_i^h = \frac{1}{n} \text{Tr}(A_n^h) = \beta_h(A_n) \text{ (say)}, \tag{2.2}$$

where Tr denotes the trace of a matrix. This is often known as the *trace-moment* formula.

Let E_{F_n} and E denote the expectations respectively, with respect to the ESD F_n, and the probability on the space where the entries of the random matrix are defined. Thus, Lemma 2.2.1 comes into force, except that now the moments are also random. Lemma 2.2.4 links convergence of moments of the ESD to those of the LSD. Consider the following conditions:

(M1) For every $h \geq 1$, $\mathrm{E}[\beta_h(A_n)] \to \beta_h$.

(M2) $\mathrm{Var}[\beta_h(A_n)] \to 0$ for every $h \geq 1$.

(M4) $\displaystyle\sum_{n=1}^{\infty} \mathrm{E}[\beta_h(A_n) - \mathrm{E}(\beta_h(A_n))]^4 < \infty$ for every $h \geq 1$.

(R) The sequence $\{\beta_h\}$ satisfies Riesz's condition (2.1).

Note that (M4) implies (M2). The following lemma follows easily from Lemma 2.2.1 and the *Borel-Cantelli Lemma*. We omit its proof.

Lemma 2.2.4. (a) If (M1), (M2) and (R) hold, then $\{F_{A_n}\}$ converges in probability to F determined by $\{\beta_h\}$.

(b) If further (M4) holds, then the convergence in (a) is almost sure.

The computation of $\mathrm{E}[\beta_h(A_n)]$ involves computation of the expected trace of A_n^h or at least its leading term. This ultimately reduces to counting the number of contributing terms in the following expansion (here a_{ij} denotes the (i,j)-th entry of A_n):

$$E[\text{Tr}(A_n^h)] = \sum_{1 \le i_1, i_2, \ldots, i_h \le n} E[a_{i_1 i_2} a_{i_2 i_3} \cdots a_{i_{h-1} i_h} a_{i_h i_1}].$$

The method is straightforward but requires all moments to be finite. This problem is usually circumvented by first assuming this to be the case, and then resorting to truncation arguments when higher moments are not necessarily finite. Also note that in specific cases, the combinatorial arguments involved may become quite unwieldy as h and n increase. Nevertheless, this approach has been successfully used to establish the existence of the LSD for several important real symmetric random matrices.

2.2.1   Scaling

The matrices need appropriate scaling for the existence of a non-trivial LSD. To understand what this scaling should be, assume that $\{x_i\}$ have mean zero and variance 1 and consider the specific case of the symmetric circulant. Let F_n denote the ESD of SC_n and let X_n denote a random variable with distribution F_n. Then

$$E_{F_n}(X_n) = \frac{1}{n} \sum_{i=1}^{n} \lambda_i = \frac{1}{n} \text{Tr}(SC_n) = x_0, \quad E[E_{F_n}(X_n)] = 0, \quad \text{and}$$

$$E_{F_n}(X_n^2) = \frac{1}{n} \sum_{i=1}^{n} \lambda_i^2 = \frac{1}{n} \text{Tr}\left(SC_n{}^2\right)$$

$$= \begin{cases} \frac{1}{n}[nx_0^2 + 2n \sum_{j=1}^{\lfloor \frac{n}{2} \rfloor} x_j^2], & \text{for } n \text{ odd,} \\ \frac{1}{n}[nx_0^2 + 2n \sum_{j=1}^{\frac{n}{2}-1} x_j^2 + x_{n/2}^2], & \text{for } n \text{ even, and} \end{cases}$$

$$E[E_{F_n}(X_n^2)] = n.$$

Hence, for stability of the second moment, it is appropriate to consider $\frac{1}{\sqrt{n}} SC_n$. The same scaling is needed for all the circulant-type matrices.

2.2.2   Input and link

The sequence of variables $\{x_i, i \ge 0\}$ that is used to construct our matrix will be called the *input sequence*. Let us begin by making the following assumption. Later we shall introduce several weaker variations of this.

Assumption 2.2.1. $\{x_i\}$ are i.i.d., $E(x_i) = 0, \text{Var}(x_i) = 1$, and $E(x_i^{2k}) < \infty$ for all $k \ge 1$.

Any of the circulant-type matrices is really constructed out of an input sequence in the following way. Let $\mathbb{Z}$ be the set of all integers and let $\mathbb{Z}_+$ denote the set of all non-negative integers. Let

$$L_n : \{1, 2, \ldots, n\}^2 \to \mathbb{Z}, \ n \geq 1 \tag{2.3}$$

be a sequence of functions. For notational convenience, we shall write $L_n = L$ and call it the *link* function. By abuse of notation we write $\mathbb{Z}_+^2$ as the common domain of $\{L_n\}$. Then each circulant-type matrix is of the form

$$A_n = ((x_{L(i,j)})), \tag{2.4}$$

and has its own distinct link function. If L is symmetric, $(L(i,j) = L(j,i)$ for all i, j), then A_n is symmetric. In particular the link functions for the symmetric circulant and the reverse circulant are respectively given by

$$L(i,j) = n/2 - |n/2 - |i - j|| \ \text{ and } \ L(i,j) = (i + j - 2) \bmod n.$$

The link functions of our circulants share some common features that shall be crucial: let $\{A_n\}$ be any sequence of circulant-type matrices. Let

$k_n = $ number of distinct variables in A_n,

$\alpha_n = $ maximum number of times a variable appears in A_n, and

$\beta_n = $ maximum number of times any variable appears in a row/column of A_n.

Then

$$k_n = O(n), \ \ \alpha_n = O(n) \ \text{ and } \ \beta_n = O(1). \tag{2.5}$$

2.2.3 Trace formula and circuits

Let $A_n = ((x_{L(i,j)}))$. Using (2.2), the h-th moment of $F_{n^{-1/2}A_n}$ is given by the following *trace formula* or the *trace-moment* formula:

$$\frac{1}{n} \operatorname{Tr} \left(\frac{A_n}{\sqrt{n}} \right)^h = \frac{1}{n^{1+h/2}} \sum_{1 \leq i_1, i_2, \ldots, i_h \leq n} x_{L(i_1, i_2)} x_{L(i_2, i_3)} \cdots x_{L(i_{h-1}, i_h)} x_{L(i_h, i_1)}.$$
$$\tag{2.6}$$

We shall now develop an equivalence relation which will help us to group the terms with the same expectation together.

Circuit: $\pi : \{0, 1, 2, \ldots, h\} \to \{1, 2, \ldots, n\}$ with $\pi(0) = \pi(h)$ is called a *circuit* of *length* h. The dependence of a circuit on h and n will be suppressed. Clearly, the convergence in (M1), (M2) and (M4) may be written in terms of circuits. For example, (M1) can be written as

$$\mathrm{E}[\beta_h(\tfrac{1}{\sqrt{n}} A_n)] = \mathrm{E}\left[\frac{1}{n} \operatorname{Tr} \left(\frac{A_n}{\sqrt{n}} \right)^h \right] = \frac{1}{n^{1+h/2}} \sum_{\pi: \ \pi \ \text{circuit}} \mathrm{E} X_\pi \to \beta_h, \tag{2.7}$$

where

$$X_\pi = x_{L(\pi(0),\pi(1))} x_{L(\pi(1),\pi(2))} \cdots x_{L(\pi(h-2),\pi(h-1))} x_{L(\pi(h-1),\pi(h))}.$$

Matched Circuit: We call a circuit π *matched* if each value $L(\pi(i-1),\pi(i))$, $1 \le i \le h$ is repeated *at least twice*. If π is non-matched, then $E(X_\pi) = 0$. If each value is repeated *exactly twice* (so h is necessarily even) then we say π is *pair-matched*, and in that case $E(X_\pi) = 1$.

To deal with (M2) or (M4), we need multiple circuits: k circuits $\pi_1, \ldots, \pi_k$ are *jointly-matched* if each L-value occurs at least twice across all circuits. They are *cross-matched* if each circuit has at least one L-value which occurs in at least one of the *other* circuits. Note that this implies that none of them are self-matched.

Equivalence of circuits: Say that π_1 and π_2 are equivalent (write $\pi_1 \sim \pi_2$) if and only if their L-values match at the *same* locations. That is, for all i, j,

$$L(\pi_1(i-1),\pi_1(i)) = L(\pi_1(j-1),\pi_1(j)) \Leftrightarrow L(\pi_2(i-1),\pi_2(i)) = L(\pi_2(j-1),\pi_2(j)). \tag{2.8}$$

This defines an equivalence relation, and if $\pi_1 \sim \pi_2$ then $E X_{\pi_1} = E X_{\pi_2}$.

2.2.4   Words and vertices

Any equivalence class can be indexed by a partition of $\{1, 2, \ldots, h\}$. Each partition block identifies the positions where the L-matches take place. We label these partitions by *words* of length h of letters where the first occurrence of each letter is in alphabetical order. For example, if $h = 5$ then the partition $\{\{1,5\}, \{2,3,4\}\}$ is represented by the word *abbba*. This identifies all circuits π for which $L(\pi(0),\pi(1)) = L(\pi(4),\pi(5))$ and $L(\pi(1),\pi(2)) = L(\pi(2),\pi(3)) = L(\pi(3),\pi(4))$. Let $w[i]$ denote the i-th entry of w.

The class Π: The equivalence class corresponding to w will be denoted by

$$\Pi(w) = \{\pi : w[i] = w[j] \Leftrightarrow L(\pi(i-1),\pi(i)) = L(\pi(j-1),\pi(j))\}.$$

The number of distinct letters in w is the number of partition blocks corresponding to w and will be denoted by $|w|$. If $\pi \in \Pi(w)$, then clearly,

$$|w| = \#\{L(\pi(i-1),\pi(i)) : 1 \le i \le h\}.$$

Note that for any fixed h, as $n \to \infty$, the number of words is finite but the number of circuits in any given $\Pi(w)$ may grow indefinitely.

Notions introduced for circuits carry over to words in an obvious way. For instance, w is *pair-matched* if every letter appears exactly twice. Let

$$\mathcal{W}_{2k}(2) = \{w : w \text{ is pair-matched and is of length } 2k\}. \tag{2.9}$$

Vertex: Any i (or $\pi(i)$ by abuse of notation) will be called a *vertex*. It is *generating*, if either $i = 0$ or $w[i]$ is the *first* occurrence of a letter. Otherwise it is called *non-generating*. For example, if $w = abbbcab$ then only $\pi(0)$, $\pi(1)$, $\pi(2)$, $\pi(5)$ are generating. Clearly for our two matrices, a circuit is completely determined, *up to a finitely many choices*, by its generating vertices. The number of generating vertices is $|w| + 1$ and hence

$$\#\Pi(w) = O(n^{|w|+1}).$$

We already know that it is enough to consider matched circuits. The next lemma shows that we can restrict attention to only pair-matched words. Let

$N = $ number of matched but not pair-matched circuits of length h.

Lemma 2.2.5. Let L be the SC_n or the RC_n link function.
(a) There is a constant C_h, depending on L and h, such that

$$N \leq C_h n^{\lfloor (h+1)/2 \rfloor} \quad \text{and hence} \quad n^{-(1+h/2)} N \to 0 \quad \text{as} \quad n \to \infty. \qquad (2.10)$$

(b) If the input sequence satisfies Assumption 2.2.1, then for every $k \geq 1$,

$$\lim_n \mathrm{E}[\beta_{2k+1}(\tfrac{1}{\sqrt{n}} A_n)] = 0 \quad \text{and} \qquad (2.11)$$

$$\lim_n \mathrm{E}[\beta_{2k}(\tfrac{1}{\sqrt{n}} A_n)] = \sum_{w \in \mathcal{W}_{2k}(2)} \lim_n \frac{1}{n^{1+k}} \#\Pi(w) \text{ (if the limit exists).} \qquad (2.12)$$

Proof. (a) Since the number of words is finite, we can fix a word and argue. Let w be a word of length h which is matched but not pair-matched. Either $h = 2k$ or $h = 2k - 1$ for some k. In both cases $|w| \leq k - 1$. If we fix the generating vertices, since $\beta_n = O(1)$ (see (2.5)), the number of choices for the non-generating vertices is upper bounded by C_h, say. Hence

$$\#\Pi(w) \leq n C_h n^{k-1} = C_h n^{\lfloor (h+1)/2 \rfloor}.$$

Relation (2.10) is an immediate consequence.

(b) For the first part of (b), since $h = 2k + 1$ is odd, there are only terms which are matched but not pair-matched. Since all moments are finite, there is a grand upper bound for all the moments. Now use part (a).

To prove the second part of (b), using the mean zero and independence assumption, (provided the last limit below exists),

$$\lim \mathrm{E}[\beta_{2k}(\tfrac{1}{\sqrt{n}} A_n)] = \lim \frac{1}{n^{1+k}} \sum_{\pi \text{ circuit}} \mathrm{E}\, X_\pi$$

$$= \sum_{w \text{ matched}} \lim \frac{1}{n^{1+k}} \sum_{\pi \in \Pi(w)} \mathrm{E}\, X_\pi. \qquad (2.13)$$

By Holder's inequality and Assumption 2.2.1, for some constant C_{2k},

$$\left| \sum_{\pi:\ \pi\in\Pi(w)} E\,X_\pi \right| \le \#\Pi(w)C_{2k}.$$

Therefore, from part (a), matched circuits which are not pair-matched do not contribute to the limit in (2.12). So

$$\lim E[\beta_{2k}(\frac{1}{\sqrt{n}}A_n)] = \sum_{w\in\mathcal{W}_{2k}(2)} \lim_n \frac{1}{n^{1+k}}\#\Pi(w), \qquad (2.14)$$

if the limit exists. This establishes (2.12) and the proof is complete. □

2.2.5 (M1) and Riesz's condition

Define, for every k and for every $w \in \mathcal{W}_{2k}(2)$,

$$p(w) = \lim_n \frac{1}{n^{1+k}}\#\Pi(w), \quad \text{whenever the limit exists.} \qquad (2.15)$$

For any fixed word, this limit will be positive and finite only if the number of elements in $\Pi(w)$ is of *exact* order n^{k+1}. Lemma 2.2.5 implies that then the limiting $(2k)$-th moment (provided the limit above exists) is the finite sum

$$\beta_{2k} = \sum_{w\in\mathcal{W}_{2k}(2)} p(w). \qquad (2.16)$$

This would essentially establish the (M1) condition. We shall verify the existence of the limit (2.15) shortly for RC_n and SC_n.

We can easily establish Riesz's condition (2.1) for these two matrices. Recall that once the generating vertices are fixed, the number of choices for each non-generating vertex is bounded above (say by Δ). [Indeed, we shall show later that for the words w for which $p(w) \ne 0$, we have $\Delta = 1$]. Hence, for each pair-matched w of length $2k$,

$$p(w) \le \Delta^k.$$

As there are $\frac{(2k)!}{k!2^k}$ pair-matched words of length $2k$, we get

$$\beta_{2k} \le \frac{(2k)!}{k!2^k}\Delta^k,$$

and hence condition (2.1) holds by Lemma 2.2.3.

2.2.6 (M4) condition

The next two lemmata help to verify (M4). We skip the proofs. For detailed proofs of these lemmata, see Chapter 1 of Bose (2018). Let

$$Q_{h,4} = \#\{(\pi_1, \pi_2, \pi_3, \pi_4) : \text{all are of length } h \text{ and are jointly- and cross-matched}\}.$$

Lemma 2.2.6. Let L be the SC_n or the RC_n link function. Then there exists a K, depending on L and h, such that

$$Q_{h,4} \leq K n^{2h+2}.$$

Lemma 2.2.7. Let $A_n = SC_n$ or RC_n. If the input sequence $\{x_i\}$ satisfies Assumption 2.2.1, then

$$E\left[\frac{1}{n} \operatorname{Tr}\left(\frac{A_n}{\sqrt{n}}\right)^h - E\frac{1}{n} \operatorname{Tr}\left(\frac{A_n}{\sqrt{n}}\right)^h\right]^4 = O(n^{-2}),$$

and hence (M4) holds.

2.3 Reverse circulant

Figure 2.1 shows the histogram of the simulated eigenvalues of $\frac{1}{\sqrt{n}} RC_n$. Bose

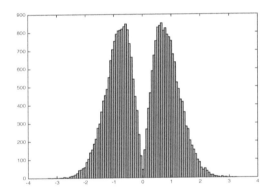

FIGURE 2.1
Histogram of the eigenvalues of $\frac{1}{\sqrt{n}} RC_n$ for $n = 1000$, 30 replications. Input is i.i.d. $N(0, 1)$.

and Mitra (2002) first established the LSD of $\frac{1}{\sqrt{n}} RC_n$ by using a normal approximation when the input variables have a finite third moment. Here we present the moment method proof from Bose and Sen (2008). The following words play a crucial role.

Definition 2.3.1. (*Symmetric word*). A pair-matched word is *symmetric* if each letter occurs once each in an odd and an even position. We shall denote the set of symmetric words of length $2k$ by $\mathcal{S}_{2k}$.

For example, $w = aabb$ is a symmetric word while $w = abab$ is not. A simple counting argument leads to the following lemma. Its proof is left as an exercise.

Lemma 2.3.1. For every $k \geq 1$, $\#\mathcal{S}_{2k} = k!$.

Let R denote a random variable which is distributed as the symmetrized square root of $\chi_2^2/2$ where χ_2^2 is a chi-squared random variable with two degrees of freedom. Let $\mathcal{L}_R$ denote its distribution. This distribution is known as the *symmetrized Rayleigh distribution*. The density and moments of this distribution are given by

$$\begin{aligned} f_R(x) &= |x| \exp(-x^2), \quad -\infty < x < \infty, \quad (2.17)\\ \beta_{2k}(\mathcal{L}_R) &= k! \text{ and } \beta_{2k+1}(\mathcal{L}_R) = 0, \quad k = 0, 1, \ldots. \end{aligned}$$

Theorem 2.3.2 (Bose and Sen (2008)). If $\{x_i\}$ satisfies Assumption 2.2.1, then the almost sure LSD of $\frac{1}{\sqrt{n}} RC_n$ is $\mathcal{L}_R$.

Proof. Using Lemmata 2.2.5, 2.2.7 and (2.17), it is enough to show that

$$\begin{aligned} \lim_n E[\beta_{2k}(\frac{1}{\sqrt{n}} RC_n)] &= \sum_{w \in \mathcal{W}_{2k}(2)} \lim_n \frac{1}{n^{k+1}} \#\Pi(w)\\ &= k!. \end{aligned}$$

Now due to Lemma 2.3.1, it is enough to verify the following two statements:

(i) If $w \in \mathcal{W}_{2k}(2) \cap \mathcal{S}_{2k}^c$ then $\lim_{n \to \infty} \frac{1}{n^{k+1}} \#\Pi(w) = 0$.

(ii) If $w \in \mathcal{S}_{2k}$ then for each choice of the generating vertices there is exactly one choice for the non-generating vertices. Hence $\lim_{n \to \infty} \frac{1}{n^{k+1}} \#\Pi(w) = 1$.

Proof of (i): It is enough to restrict attention to pair-matched words. Let

$$v_i = \pi(i)/n \text{ and } t_i = v_i + v_{i-1}.$$

Note that the vertices i, j match if and only if

$$\begin{aligned} (\pi(i-1) + \pi(i) - 2) \bmod n &= (\pi(j-1) + \pi(j) - 2) \bmod n\\ \Leftrightarrow t_i - t_j &= 0, 1 \text{ or } -1. \end{aligned}$$

Since w is pair-matched, let $\{(i_s, j_s), 1 \leq s \leq k\}$ be such that $w[i_s] = w[j_s]$, where $j_s, 1 \leq s \leq k$, is in ascending order and $j_k = 2k$. Define $U_n = \{0, \frac{1}{n}, \frac{2}{n}, \ldots \frac{n-1}{n}\}$. Let $r = (r_1, \ldots, r_k)$ denote a typical sequence in $\{0, \pm 1\}^k$. Then we can write

$$\#\Pi(w) = \sum_r \#\{(v_0, v_1, \ldots, v_{2k}) : v_0 = v_{2k}, \ v_i \in U_n, \text{ and } t_{i_s} - t_{j_s} = r_s\}.$$

Let S be the set of generating vertices and $v_S = \{v_i : i \in S\}$. Any $v_i, i \notin S$ is a linear combination of the elements in v_S and we can write

$$v_i = L_i(v_S) + a_i^{(r)}, \ (i \notin S),$$

for some integer $a_i^{(r)}$. Clearly, $\frac{1}{n^{k+1}}\#\Pi(w)$ can be written as a Riemann sum (in $k+1$ dimension) and it converges to

$$\sum_r \underbrace{\int_0^1 \cdots \int_0^1}_{k+1} \mathbb{I}(0 \leq L_i(v_S)+a_i^{(r)} \leq 1, \, \forall\, i \notin S\cup\{2k\})\mathbb{I}(v_0 = L_{2k}(v_S)+a_{2k}^{(r)})dv_S.$$

$$(2.18)$$

The first indicator appears since every non-generating vertex i, when solved in terms of the generating vertices, must lie in U_n. The second indicator comes from the fact that the numerical value of the non-generating vertex $2k$ equals the numerical value of the vertex 0 due to the circuit condition.

Now assume that this limit is non-zero. Then at least one of the terms in the above sum must be non-zero. This automatically implies that we must have (otherwise it is the volume of a subspace in dimension k or less and hence is zero)

$$v_0 = L_{2k}(v_S) + a_{2k}^{(r)}. \tag{2.19}$$

Now $(t_{i_s} - t_{j_s} - r_s) = 0$ for all $s, 1 \leq s \leq k$. Hence trivially, for any choice of $\{\alpha_s\}$,

$$v_{2k} = v_{2k} + \sum_{s=1}^k \alpha_s(t_{i_s} - t_{j_s} - r_s).$$

Let us choose integers $\{\alpha_s\}$ as follows: let $\alpha_k = 1$. Having fixed $\alpha_k, \alpha_{k-1}, \ldots,$ α_{s+1}, we choose α_s as follows:

(a) if $j_s + 1 \in \{i_m, j_m\}$ for some $m > s$, then set $\alpha_s = \pm\alpha_m$ according as $j_s + 1$ equals i_m or j_m,

(b) if there is no such m, choose α_s to be any integer.

By this choice of $\{\alpha_s\}$, we ensure that in $v_{2k} + \sum_{s=1}^k \alpha_s(t_{i_s} - t_{j_s} - r_s)$, the coefficients of each $v_i, i \notin S$ cancel out. Hence we get

$$v_{2k} = v_{2k} + \sum_{s=1}^k \alpha_s(t_{i_s} - t_{j_s} - r_s) = L(v_S) + a \quad \text{(some linear combination)}.$$

However, from (2.19), $v_0 = L_{2k}^H(v_S) + a_{2k}^{(r)}$. Hence, because only generating vertices are left in both the linear combinations,

$$v_{2k} + \sum_{s=1}^k \alpha_s(t_{i_s} - t_{j_s} - r_s) - v_0 = 0, \tag{2.20}$$

and thus the coefficient of each v_i in the left side has to be zero including the constant term.

Now consider the coefficients of $\{t_i\}$ in (2.20). First, since $\alpha_k = 1$, the coefficient of t_{2k} is -1. On the other hand, the coefficient of v_{2k-1} is 0. Hence the coefficient of t_{2k-1} has to be $+1$.

Proceeding to the next step, we know that the coefficient of v_{2k-2} is 0. However, we have just observed that the coefficient of t_{2k-1} is +1. Hence the coefficient of t_{2k-2} must be -1. If we continue in this manner, in the expression (2.20) for all odd i, t_i must have coefficient +1 and for all even i, t_i must have coefficient -1.

Now suppose that for some s, i_s and j_s both are odd or both are even. Then for any choice of α_s, t_{i_s} and t_{j_s} will have opposite signs in the expression (2.20). This contradicts the fact stated in the previous paragraph. Hence, either i_s is odd and j_s is even, or the other way around. Since this happens for all $s, 1 \leq s \leq k$, w must be a symmetric word, proving (i).

Proof of (ii): Let $w \in S_{2k}$. First fix the generating vertices. Then we determine the non-generating vertices from left to right. Consider $L(\pi(i-1), \pi(i)) = L(\pi(j-1), \pi(j))$, where $i < j$ and $\pi(i-1)$, $\pi(i)$ and $\pi(j-1)$ have been determined. We rewrite it as

$$\pi(j) = Z + dn \text{ for some integer } d \text{ where } Z = \pi(i-1) + \pi(i) - \pi(j-1).$$

Clearly $\pi(j)$ can be determined uniquely from the above equation since $1 \leq \pi(j) \leq n$. Continuing, we obtain the whole circuit uniquely. Hence the first part of (ii) is proven.

As a consequence, for $w \in S_{2k}$, only one term in the sum (2.18) will be non-zero and that term equals 1. Since there are exactly $k!$ symmetric words, (ii) is proven completely. This completes the proof of the theorem. $\qquad\square$

2.4   Symmetric circulant

A histogram of the simulated eigenvalues of $\frac{1}{\sqrt{n}} SC_n$ given in Figure 2.2 suggests that its LSD is normal. We now state this as a theorem and pro-

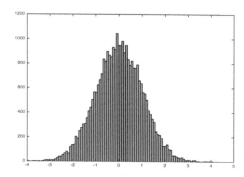

FIGURE 2.2
Histogram of the eigenvalues of $\frac{1}{\sqrt{n}} SC_n$ for $n = 1000$, 30 replications. Input is i.i.d. $N(0, 1)$.

vide a moment method proof given originally by Bose and Sen (2008). The details are similar to, but much simpler than, the proof of Theorem 2.3.2.

Theorem 2.4.1 (Bose and Sen (2008)). If $\{x_i\}$ satisfies Assumption 2.2.1, then the almost sure LSD of $\frac{1}{\sqrt{n}}SC_n$ is standard normal.

Proof. Using Lemmata 2.2.5 and 2.2.7, it suffices to show that the even moments converge to the moments of a normal random variable. The $(2k)$-th moment of the standard normal is given by $\frac{(2k)!}{2^k k!}$. Since

$$\#\mathcal{W}_{2k}(2) = \frac{(2k)!}{2^k k!},$$

it is thus enough to show that

$$\lim_{n \to \infty} \frac{1}{n^{k+1}} \#\Pi(w) = 1, \ \forall \ w \in \mathcal{W}_{2k}(2).$$

Let the *slopes* be defined by

$$s(l) = \pi(l) - \pi(l-1).$$

Clearly, vertices i, j, $i < j$, match if and only if

$$\begin{aligned} |n/2 - |s_i|| &= |n/2 - |s_j|| \\ \Leftrightarrow s(i) - s(j) &= 0, \pm n \quad \text{Or} \quad s(i) + s(j) = 0, \pm n. \end{aligned}$$

That is six possibilities in all. We first show that the first three possibilities do not contribute asymptotically.

Lemma 2.4.2. Fix a pair-matched word w with $|w| = k$. Let N be the number of pair-matched circuits of w which have at least one pair $i < j$, $s(i) - s(j) = 0, \pm n$. Then, as $n \to \infty$, $N = O(n^k)$ and hence $n^{-(k+1)}N \to 0$.

Proof. Let $(i_1, j_1), (i_2, j_2), \ldots, (i_k, j_k)$ denote the pair-partition corresponding to the word w, so that $w[i_l] = w[j_l], 1 \le l \le k$ and $i_1 < i_2 < \cdots < i_k$. Suppose, without loss of generality,

$$s(i_k) - s(j_k) = 0, \pm n. \tag{2.21}$$

Clearly a circuit π becomes completely specified if we know $\pi(0)$ and all the $\{s(i)\}$.

As already observed, if we fix some value for $s(i_l)$, there are at most six options for $s(j_l)$. We may choose the values of $\pi(0), s(i_1), s(i_2), \ldots, s(i_{k-1})$ in $O(n^k)$ ways and then we may choose values of $s(j_1), s(j_2), \ldots, s(j_{k-1})$ in $O(6^k)$ ways. For any such choice, from the sum restriction

$$\sum_{i=1}^{2k} s(i) = \pi(2k) - \pi(0) = 0,$$

we know $s(i_k) + s(j_k)$. On the other hand, equation (2.21) holds. As a consequence, the pair $(s(i_k), s(j_k))$ has at most 3 possibilities. This implies that there are at most $O(n^k)$ circuits with the given restrictions and the proof of the lemma is complete. □

Now we continue with the proof of Theorem 2.4.1. Due to the above lemma, it now remains to show that with

$$\Pi'(w) = \{\pi : \pi \text{ is a circuit}, w[i] = w[j] \Rightarrow s_i + s_j = 0, \pm n\},$$

$$\lim_{n \to \infty} \frac{1}{n^{k+1}} \#\Pi'(w) = 1.$$

Suppose for some $i < j$,

$$s_i + s_j = 0, \pm n.$$

Reading left to right, if we know the circuit up to position $(j - 1)$ then $\pi(j)$ has to take one of the values $A - n, A, A + n$, where

$$A = \pi(j - 1) - \pi(i) + \pi(i - 1).$$

Noting that

$$-(n - 2) \leq A \leq (2n - 1),$$

exactly one of the three values will fall within 1 and n and hence be a valid choice for $\pi(j)$. Thus we first choose the generating vertices arbitrarily, then the non-generating vertices are determined, from left to right uniquely, so that

$$s(i) + s(j) = 0, \pm n.$$

This automatically yields $\pi(0) = \pi(2k)$ as follows:

$$\pi(2k) - \pi(0) = \sum_{i=1}^{2k} s_i = dn \text{ for some } d \in \mathbb{Z}.$$

But since $|\pi(2k) - \pi(0)| \leq n - 1$, we must have $d = 0$. Thus $\#\Pi'(w) = n^{k+1}$, and hence $\lim_{n \to \infty} \frac{1}{n^{k+1}} \#\Pi'(w) = 1$, proving the theorem. □

2.5   Related matrices

For the symmetric Toeplitz matrix, $L(i, j) = |i - j|$ and for the symmetric Hankel matrix, $L(i, j) = i + j$. The symmetric circulant SC_n may be considered as a *doubly symmetric Toeplitz matrix*. In a similar manner the *doubly symmetric Hankel matrix* DH_n is defined as

$$DH_n = \begin{bmatrix} x_0 & x_1 & x_2 & \cdots & x_3 & x_2 & x_1 \\ x_1 & x_2 & x_3 & \cdots & x_2 & x_1 & x_0 \\ x_2 & x_3 & x_4 & \cdots & x_1 & x_0 & x_1 \\ & & & \vdots & & & \\ x_2 & x_1 & x_0 & \cdots & x_5 & x_4 & x_3 \\ x_1 & x_0 & x_1 & \cdots & x_4 & x_3 & x_2 \end{bmatrix}.$$

Its link function is

$$L(i,j) = n/2 - |n/2 - ((i+j-2) \bmod n)|, \ 1 \le i, j \le n.$$

Massey et al. (2007) defined a (symmetric) matrix to be *palindromic* if its first row is a palindrome. The palindromic Toeplitz PT_n is defined below. PH_n is defined similarly.

$$PT_n = \begin{bmatrix} x_0 & x_1 & x_2 & \cdots & x_2 & x_1 & x_0 \\ x_1 & x_0 & x_1 & \cdots & x_3 & x_2 & x_1 \\ x_2 & x_1 & x_0 & \cdots & x_4 & x_3 & x_2 \\ & & & \vdots & & & \\ x_1 & x_2 & x_3 & \cdots & x_1 & x_0 & x_1 \\ x_0 & x_1 & x_2 & \cdots & x_2 & x_1 & x_0 \end{bmatrix}.$$

We outline in the exercises how the LSD of all these matrices can be derived from the symmetric circulant LSD.

2.6 Reduced moment

The "all moments finite" stipulation in Assumption 2.2.1 is clearly restrictive. We now show how this condition can be significantly relaxed. Again, the fact that we are dealing with real symmetric matrices helps us. Our goal in this section is to show that all LSD results in this chapter remain valid when the input sequence satisfies the following assumption.

Assumption 2.6.1. $\{x_i\}$ are i.i.d. with mean zero and variance 1.

2.6.1 A metric

It is well-known that weak convergence of probability measures is metrizable. We work with a specific metric which is defined on the space of all probability distributions with finite second moment. The W_2-metric is defined on this space as follows. The distance between any two distribution functions F and G with finite second moment is defined as

$$W_2(F, G) = \left[\inf_{(X \sim F, Y \sim G)} E[X - Y]^2 \right]^{\frac{1}{2}}.$$

Here the infimum is taken over all pairs of random variables (X, Y) such that their marginal distributions are F and G, respectively. The following lemma links weak convergence and convergence in the above metric. We omit its proof which can be found in Bose (2018).

Lemma 2.6.1. W_2 is a complete metric and $W_2(F_n, F) \to 0$ if and only if $F_n \overset{\mathcal{D}}{\to} F$ and $\beta_2(F_n) \to \beta_2(F)$.

An estimate of the metric distance W_2 between two ESDs in terms of the trace will be crucial to us.

Lemma 2.6.2. Suppose A, B are $n \times n$ real symmetric matrices with eigenvalues $\{\lambda_1(A) \leq \cdots \leq \lambda_n(A)\}$ and $\{\lambda_1(B) \leq \cdots \leq \lambda_n(B)\}$. Then

$$W_2^2(F_A, F_B) \leq \frac{1}{n} \sum_{i=1}^{n} (\lambda_i(A) - \lambda_i(B))^2 \leq \frac{1}{n} \operatorname{Tr}(A - B)^2.$$

Proof. The first inequality follows by considering the joint distribution which puts mass $1/n$ at $(\lambda_i(A), \lambda_i(B))$. Then the marginals are the two ESDs of A and B. The second inequality follows from the *Hoffmann-Wielandt inequality* (see Hoffman and Wielandt (1953)). ☐

2.6.2 Minimal condition

Theorem 2.6.3 (Bose and Sen (2008)). Suppose the input sequence satisfies Assumption 2.6.1. Then the LSDs of $\frac{1}{\sqrt{n}} RC_n$ and $\frac{1}{\sqrt{n}} SC_n$ continue to be as before almost surely.

Proof. We shall briefly sketch the proof. The reader is invited to complete the details. First consider the SC_n. Fix a level of truncation $K > 0$. Define the variables

$$x_{i,K} = \sigma_K^{-1} \left[x_i I(|x_i| \leq K) - E\{x_i I(|x_i| \leq K)\} \right],$$

where

$$\sigma_K^2 = \operatorname{Var}[x_i I(|x_i| \leq K)].$$

Then $\{x_{i,K}\}$ are mean zero variance 1 i.i.d. random variables. For sufficiently large K, σ_K^2 is bounded away from zero and hence $\{x_{i,K}\}$ are also bounded. Moreover $\sigma_K^2 \to 1$ as $K \to \infty$. Let $SC_{n,K}$ be the symmetric circulant with the input sequence $\{x_{i,K}\}$. By Theorem 2.4.1 the almost sure LSD of $\frac{1}{\sqrt{n}} SC_{n,K}$ is standard normal.

On the other hand,

$$W_2^2(F_{n^{-1/2}SC_n}, F_{n^{-1/2}SC_{n,K}}) \leq \frac{1}{n^2} \text{Tr}(SC_n - SC_{n,K})^2$$

$$= \frac{1}{n}\sum_{i=0}^{n-1}[x_i - x_{i,K}]^2$$

$$\rightarrow \quad \text{E}[x_1 - x_{1,K}]^2 \text{ almost surely as } n \rightarrow \infty.$$

Now if we let $K \rightarrow \infty$, it is easy to show by an application of *Dominated Convergence Theorem* (DCT) that the right side converges to zero. The proof is then essentially complete. The same proof works for the reverse circulant. $\square$

2.7 Exercises

1. Show that in Definition 2.1.3 of weak convergence of random probability measures, $(i) \Rightarrow (ii)$.

2. Prove Lemma 2.2.1.

3. Suppose $\{X_i\}$ are i.i.d. with mean zero and variance 1 and with all moments finite. Using Lemma 2.2.1, show that $\frac{1}{\sqrt{n}}(X_1 + \cdots + X_n)$ converges weakly to the standard normal distribution. Then use the W_2 metric to relax the all moment finite condition.

4. Prove Lemma 2.2.2.

5. Prove Lemma 2.2.3.

6. Prove Lemma 2.2.4.

7. Prove Lemma 2.2.6.

8. Prove Lemma 2.2.7.

9. Prove Lemma 2.3.1.

10. Show that
    (a) the $n \times n$ principal minor of DH_{n+3} is PH_n.
    (b) the $n \times n$ principal minor of SC_{n+1} is PT_n.

11. Let J_n be the $n \times n$ matrix with entries 1 in the main anti-diagonal and zero elsewhere. Show that

    $$(PH_n)J_n = J_n(PH_n) = PT_n \quad \text{and} \quad (PT_n)^{2k} = (PH_n)^{2k}.$$

12. Show that under Assumption 2.2.1, the almost sure LSD of all the

three matrices $\frac{1}{\sqrt{n}}PT_n$, $\frac{1}{\sqrt{n}}PH_n$ and $\frac{1}{\sqrt{n}}DH_n$ is the standard normal N with $\beta_{2k}(N) = \frac{(2k)!}{2^k k!}$. Hint: Use the previous two exercises and the *interlacing inequality* (see Bhatia (1997)).

13. Prove the first inequality in Lemma 2.6.2.

14. Complete the details in the proof of Theorem 2.6.3.

15. Show that under Assumption 2.6.1, the almost sure LSD of all the three matrices $\frac{1}{\sqrt{n}}PT_n$, $\frac{1}{\sqrt{n}}PH_n$ and $\frac{1}{\sqrt{n}}DH_n$ is standard normal.

16. Simulate the eigenvalues of $\frac{1}{\sqrt{n}}RC_n$ for $n = 1000$ when the input sequence is i.i.d. standardized $U(0,1)$, and construct a histogram plot to verify Figure 2.1.

17. Simulate the eigenvalues of $\frac{1}{\sqrt{n}}SC_n$ for $n = 1000$ when the input sequence is i.i.d. standardized $U(0,1)$, and obtain a histogram plot to verify Figure 2.2.

3

LSD: normal approximation

In Chapter 1, we have derived explicit formulae for the eigenvalues of circulant-type matrices. This makes the method of normal approximation ideally suited for studying their LSD. In this chapter we show how normal approximation can be used to establish the LSD of circulant-type random matrices, symmetric as well as non-symmetric, with independent entries. The case of dependent entries will be treated in the next chapter. In the later chapters, more sophisticated normal approximation results will be used to study extreme eigenvalues and spectral gaps.

3.1 Method of normal approximation

In Chapter 2 we have established the almost sure LSD of the symmetric matrices SC_n and RC_n when the second moment is finite. We are now going to encounter non-symmetric matrices and we are also going to drop the identically distributed assumption. Hence we strengthen the moment assumption on the entries as follows:

Assumption 3.1.1. $\{x_i\}$ are independent, $\mathrm{E}(x_i) = 0$, $\mathrm{Var}(x_i) = 1$, and $\sup_i \mathrm{E}\,|x_i|^{2+\delta} < \infty$ for some $\delta > 0$.

Instead of aiming for almost sure convergence, we shall be satisfied with the weaker L_2-convergence. Recall that the ESD of $\{A_n\}$ converges to the distribution function F in L_2 if, at all continuity points (x, y) of F,

$$\int_\Omega \left[F_{A_n}(x, y) - F(x, y) \right]^2 d\mathrm{P}(\omega) \to 0 \text{ as } n \to \infty. \tag{3.1}$$

Note that the above relation holds if at all continuity points (x, y) of F,

$$\mathrm{E}[F_{A_n}(x, y)] \to F(x, y) \text{ and } \mathrm{Var}[F_{A_n}(x, y)] \to 0. \tag{3.2}$$

We often write F_n for F_{A_n} when the sequence of matrices under consideration is clear from the context.

We shall use the following result on normal approximation (Berry-Esseen bound). Its proof follows easily from Corollary 18.1, page 181 and Corollary 18.3, page 184 of Bhattacharya and Ranga Rao (1976).

Lemma 3.1.1. Let $X_1, \ldots, X_k$ be independent random vectors with values in $\mathbb{R}^d$, having zero means and an average positive-definite covariance matrix $V_k = k^{-1} \sum_{j=1}^{k} \text{Cov}(X_j)$. Let G_k denote the distribution of $k^{-1/2} T_k (X_1 + \cdots + X_k)$, where T_k is the symmetric, positive-definite matrix which satisfies $T_k^2 = V_k^{-1}$, $n \geq 1$. If $\text{E} \|X_j\|^{(2+\delta)} < \infty$ for some $\delta > 0$, then there exists $C > 0$ (depending only on d), such that

(a)

$$\sup_{B \in \mathcal{C}} |G_k(B) - \Phi_d(B)| \leq C k^{-\delta/2} [\lambda_{\min}(V_k)]^{-(2+\delta)} \rho_{2+\delta};$$

(b) for any Borel set A,

$$|G_k(A) - \Phi_d(A)| \leq C k^{-\delta/2} [\lambda_{\min}(V_k)]^{-(2+\delta)} \rho_{2+\delta} + 2 \sup_{y \in \mathbb{R}^d} \Phi_d((\partial A)^\eta - y),$$

where Φ_d is the d-dimensional standard normal distribution function, $\mathcal{C}$ is the class of all Borel measurable *convex* subsets of $\mathbb{R}^d$,

$$\rho_{2+\delta} = k^{-1} \sum_{j=1}^{k} \text{E} \|X_j\|^{(2+\delta)} \quad \text{and} \quad \eta = C \rho_{2+\delta} n^{-\delta/2}.$$

3.2   Circulant

The first theorem is on the LSD of C_n with independent input. Figure 3.1 provides a scatter plot of the eigenvalues of this matrix for $n = 2000$.

Theorem 3.2.1. If Assumption 3.1.1 is satisfied then the ESD of $\frac{1}{\sqrt{n}} C_n$ converges in L_2 to the two-dimensional normal distribution given by $\mathbf{N}(0, D)$, where D is a 2×2 diagonal matrix with diagonal entries $1/2$.

Remark 3.2.1. Sen (2006) had proven the above result under a third moment assumption. Meckes (2009) established the result for independent complex entries that satisfies $\text{E}(x_j) = 0$, $\text{E} |x_j|^2 = 1$ and for every $\epsilon > 0$,

$$\lim_{n \to \infty} \frac{1}{n} \sum_{j=0}^{n-1} \text{E}(|x_j|^2 \mathbb{I}_{\{|x_j| > \epsilon \sqrt{n}\}}) = 0. \tag{3.3}$$

This generality is obtained by using Lindeberg's central limit theorem along with the normal approximation bounds. The proof is left as an exercise.

Proof of Theorem 3.2.1. First recall the eigenvalues of C_n from (1.2) of Section 1.1. Then observe that we may ignore the eigenvalue λ_n and also $\lambda_{n/2}$

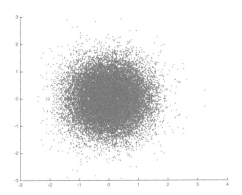

FIGURE 3.1
Eigenvalues of $\frac{1}{\sqrt{n}}C_n$ for $n = 2000$, 10 replications. Input is i.i.d. $N(0, 1)$.

whenever n is even, since they contribute at most $2/n$ to the ESD $F_n(x, y)$. So for $x, y \in \mathbb{R}$,

$$E[F_n(x, y)] \sim \frac{1}{n} \sum_{k=1, k \neq n/2}^{n-1} P(b_k \leq x, c_k \leq y),$$

where

$$b_k = \frac{1}{\sqrt{n}} \sum_{j=0}^{n-1} x_j \cos(\omega_k j), \quad c_k = \frac{1}{\sqrt{n}} \sum_{j=0}^{n-1} x_j \sin(\omega_k j), \text{ and } \omega_k = \frac{2\pi k}{n}.$$

Recall from (3.2) that it is enough to show that for all $x, y \in \mathbb{R}$,

$$E[F_n(x, y)] \to \Phi_{0,D}(x, y) \quad \text{and} \quad \text{Var}[F_n(x, y)] \to 0.$$

To show $E[F_n(x, y)] \to \Phi_{0,D}(x, y)$, define for $1 \leq l, k \leq n - 1$, $k \neq n/2$,

$$X_{l,k} = \left(\sqrt{2}x_l \cos(\omega_k l), \quad \sqrt{2}x_l \sin(\omega_k l)\right)'.$$

Note that

$$E(X_{l,k}) = 0, \quad \frac{1}{n} \sum_{l=0}^{n-1} \text{Cov}(X_{l,k}) = I, \text{ and} \qquad (3.4)$$

$$\sup_n \sup_{1 \leq k \leq n} \left[\frac{1}{n} \sum_{l=0}^{n-1} E \parallel X_{lk} \parallel^{(2+\delta)}\right] \leq C < \infty. \qquad (3.5)$$

For $k \neq n/2$,

$$(b_k \leq x, c_k \leq y) = \left\{\frac{1}{\sqrt{n}} \sum_{l=0}^{n-1} X_{l,k} \leq (\sqrt{2}x, \sqrt{2}y)'\right\}.$$

Since $\{(r, s) : (r, s)' \leq (\sqrt{2}x, \sqrt{2}y)'\}$ is *convex* in $\mathbb{R}^2$ and $\{X_{l,k},\ 0 \leq l \leq n-1\}$ satisfies (3.5), we can apply Lemma 3.1.1(a) for $k \neq n/2$ to get,

$$\left| P\left(\frac{1}{\sqrt{n}} \sum_{l=0}^{n-1} X_{l,k}\right) \leq (\sqrt{2}x, \sqrt{2}y)' \right) - P\left((N_1,\ N_2)' \leq (\sqrt{2}x, \sqrt{2}y)'\right) \right|$$

$$\leq Cn^{-\delta/2}\left[\frac{1}{n} \sum_{l=0}^{n-1} E\,\|X_{lk}\|^{(2+\delta)}\right] \leq Cn^{-\delta/2} \to 0, \text{ as } n \to \infty,$$

where N_1 and N_2 are i.i.d. standard normal variables. Therefore

$$\lim_{n\to\infty} E[F_n(x, y)] = \lim_{n\to\infty} \frac{1}{n} \sum_{k=1, k\neq n/2}^{n-1} P\left(b_k \leq x, c_k \leq y\right)$$

$$= \lim_{n\to\infty} \frac{1}{n} \sum_{k=1, k\neq n/2}^{n-1} P\left((N_1,\ N_2)' \leq (\sqrt{2}x, \sqrt{2}y)'\right)$$

$$= \Phi_{0,D}(x, y). \tag{3.6}$$

Now, to show $\mathrm{Var}[F_n(x, y)] \to 0$, it is enough to show that

$$\frac{1}{n^2} \sum_{k\neq k'=1}^{n} \mathrm{Cov}(J_k, J_{k'}) = \frac{1}{n^2} \sum_{k\neq k'=1}^{n} [E(J_k J_{k'}) - E(J_k)\,E(J_{k'})] \to 0, \quad (3.7)$$

where J_k is the indicator that $\{b_k \leq x, c_k \leq y\}$. Now as $n \to \infty$,

$$\frac{1}{n^2} \sum_{k\neq k'=1}^{n} E(J_k)\,E(J_{k'}) = \left[\frac{1}{n} \sum_{k=1}^{n} E(J_k)\right]^2 - \frac{1}{n^2} \sum_{k=1}^{n} \left[E(J_k)\right]^2$$

$$\to \left[\Phi_{0,D}(x, y)\right]^2.$$

So to show (3.7), it is enough to show that as $n \to \infty$,

$$\frac{1}{n^2} \sum_{k\neq k'; k, k'=1}^{n} E(J_k, J_{k'}) \to \left[\Phi_{0,D}(x, y)\right]^2.$$

Along the lines of the proof used to show (3.6), one may now extend the vectors of two coordinates defined above to ones with four coordinates, and proceed exactly as above to verify this. We omit the routine details. This completes the proof of Theorem 3.2.1. $\qquad\square$

Using the ideas in the above proof, it is not hard to prove the following result for SC_n and RC_n with independent entries. We leave it as an exercise.

Theorem 3.2.2. Suppose $\{x_i\}$ satisfies Assumption 3.1.1. Then

(a) the ESD of $\frac{1}{\sqrt{n}}SC_n$ converges in L_2 to the standard normal distribution, and

(b) the ESD of $\frac{1}{\sqrt{n}}RC_n$ converges in L_2 to the symmetrized Rayleigh distribution $\mathcal{L}_R$ defined in (2.17).

3.3 *k*-circulant

From the formula solution of the eigenvalues of the *k*-circulant matrix given in Theorem 1.4.1, it is clear that for many combinations of k and n, a lot of eigenvalues are zero. For example, if k is prime and $n = m \times k$ where $\gcd(m, k) = 1$, then 0 is an eigenvalue with multiplicity $(n - m)$. To avoid this degeneracy and to keep our exposition simple, we primarily restrict our attention to the case when $\gcd(k, n) = 1$.

In general, the structure of the eigenvalues depends on the relation between k and n. For any fixed value of k other than 1, the LSD starts to depend on the particular subsequence along which $n \to \infty$. For example, if $k = 3$, then the behavior of the ESD depends on whether n is or is not a multiple of 3 (see Figure 3.2). See Bose (2018) for more such simulation examples.

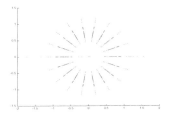 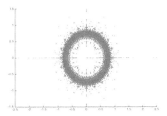

FIGURE 3.2

Eigenvalues of 20 realizations of $\frac{1}{\sqrt{n}} A_{k,n}$ with input i.i.d. $N(0,1)$ when (i) (left) $k = 3$, $n = 999$ and (ii) (right) $k = 3$, $n = 1000$.

The next theorem, due to Bose et al. (2012b), implies that the radial component of the LSD of *k*-circulants with $k \geq 2$ is always degenerate, at least when the input sequence is i.i.d. normal, as long as $k = n^{o(1)}$ and $\gcd(k, n) = 1$. Observe that, in this case also n tends to infinity along a sub-sequence, so that the condition $\gcd(k, n) = 1$ continues to hold.

Theorem 3.3.1 (Bose et al. (2012b)). Suppose $\{x_i\}_{i \geq 0}$ is an i.i.d. sequence of $N(0, 1)$ random variables. Let $k \geq 2$ be such that $k = n^{o(1)}$ and $n \to \infty$ with $\gcd(n, k) = 1$. Then $F_{n^{-1/2} A_{k,n}}$ converges weakly in probability to the uniform distribution over the circle with center at $(0, 0)$ and radius $r = \exp(\mathrm{E}\left[\log \sqrt{E}\right])$, E being an exponential random variable with mean one.

Remark 3.3.1. The random variable $-\log E$ has the standard Gumbel distribution with mean $\gamma = \lim\limits_{n \to \infty} \left[1 + \frac{1}{2} + \cdots + \frac{1}{n} - \log n\right] \approx 0.57721$ (the Euler-Mascheroni constant). It follows that $r = e^{-\gamma/2} \approx 0.74930$.

It is thus natural to consider the case when k^g is of the order n and $\gcd(k, n) = 1$ where g is a fixed integer. In the next two theorems, we consider two special cases of the above scenario, namely when n divides $k^g \pm 1$.

We shall need the following lemma to prove Theorem 3.3.1. We shall also use it in Chapters 6 and 8.

Lemma 3.3.2. Fix k and n. Suppose that $\{x_l\}_{0 \leq l < n}$ are i.i.d. standard normal random variables. Recall the notation $\mathcal{P}_j$, n_j and y_j from (1.7) and (1.8) in Chapter 1. Then

(a) For every n, $\frac{1}{\sqrt{n}}a_{t,n}$, $\frac{1}{\sqrt{n}}b_{t,n}$, $0 \leq t \leq n/2$ are i.i.d. normal with mean zero and variance $1/2$, and any sub-collection $\{y_{j_1}, y_{j_2}, \ldots\}$ of $\{y_j\}_{0 \leq j < \ell}$, so that none of them are conjugates of each other, are mutually independent where

$$a_{t,n} = \sum_{j=0}^{n-1} x_j \cos\left(\frac{2\pi t j}{n}\right) \text{ and } b_{t,n} = \sum_{j=0}^{n-1} x_j \sin\left(\frac{2\pi t j}{n}\right).$$

(b) If $1 \leq j < \ell$ where $\mathcal{P}_j \cap (n - \mathcal{P}_j) = \emptyset$, then $n^{-n_j/2} y_j$ is distributed as the product of n_j i.i.d. copies of $E^{1/2}U$ where E and U are independent, E is exponential with mean one, and U is uniform over the unit circle in $\mathbb{R}^2$.

(c) Suppose $1 \leq j < \ell$ and $\mathcal{P}_j = n - \mathcal{P}_j$ and $n/2 \notin \mathcal{P}_j$. Then $n^{-n_j/2}\Pi_j$ are distributed as the product of $(n_j/2)$-fold product of i.i.d. exponentials with mean one.

Proof. (a) Proof of this part is straight forward, so we skip it.

(b) By part (a), $\frac{1}{\sqrt{n}}\lambda_t = \frac{1}{\sqrt{n}}a_{t,n} + i\frac{1}{\sqrt{n}}b_{t,n}$ is a complex normal random variable with mean zero and variance $1/2$ for every $0 < t < n$. Moreover, they are independent by the given restriction on $\mathcal{P}_j$. Now observe that such a complex normal is distributed as $E^{1/2}U$.

(c) If $t \in \mathcal{P}_j$ then $n - t \in \mathcal{P}_j$ too and $t \neq n - t$. By part (a), $\frac{1}{n}\lambda_t\lambda_{n-t} = \frac{1}{n}\left(a_{t,n}^2 + b_{t,n}^2\right)$ which is distributed as $Y/2$ where Y is Chi-square with two degrees of freedom. This is the same distribution as that of an exponential random variable with mean one. The proof is complete once we observe that n_j is necessarily even and the λ_t's associated with $\mathcal{P}_j$ can be grouped into $n_j/2$ disjoint pairs like above, which are mutually independent. $\square$

Proof of Theorem 3.3.1. Recall $\lambda_j, \mathcal{P}_j, n_j$ and g_x from Section 1.4. By Theorem 1.4.1, the eigenvalues of $\frac{1}{\sqrt{n}}A_{k,n}$ are

$$\exp\left(\frac{2\pi i(s + \Theta_j)}{n_j}\right) \times \left(\prod_{t \in \mathcal{P}_j} |\frac{1}{\sqrt{n}}\lambda_t|\right)^{1/n_j}, \quad 1 \leq s \leq n_j, \ 0 \leq j < \ell,$$

where $2\pi\Theta_j = \arg(\prod_{t \in \mathcal{P}_j} \lambda_t), \Theta_j \in [0,1)$ and $\arg(z)$ is the usual argument of z between 0 and 2π. Fix any $\epsilon > 0$ and $0 < \theta_1 < \theta_2 < 2\pi$. Define

$$B(\theta_1, \theta_2, \epsilon) = \{(x, y) \in \mathbb{R}^2 : r - \epsilon < \sqrt{x^2 + y^2} < r + \epsilon, \tan^{-1}(y/x) \in [\theta_1, \theta_2]\}.$$

Clearly, it is enough to prove that as $n \to \infty$,

$$\frac{1}{n} \sum_{j=0}^{\ell-1} \sum_{s=1}^{n_j} \mathbb{I}\left(\exp\left(\frac{2\pi i(s + \Theta_j)}{n_j} \right) \times \left(\prod_{t \in \mathcal{P}_j} |\frac{\lambda_t}{\sqrt{n}}| \right)^{1/n_j} \in B(\theta_1, \theta_2, \epsilon) \right) \xrightarrow{\mathcal{P}} \frac{\theta_2 - \theta_1}{2\pi}.$$

(3.8)

Note that for a fixed positive integer C, we have

$$\frac{1}{n} \sum_{1 \le j < \ell : n_j \le C} n_j \le \frac{1}{n} \sum_{u=2}^{C} \#\{1 \le x < n : g_x = u\}$$

$$\le \frac{1}{n} \sum_{u=2}^{C} \#\{1 \le x < n : xk^u = x \bmod n\}$$

$$= \frac{1}{n} \sum_{u=2}^{C} \#\{1 \le x < n : x(k^u - 1) = sn \text{ for some } s \ge 1\}$$

$$\le \frac{1}{n} \sum_{u=2}^{C} (k^u - 1) \le \frac{1}{n} Ck^C \to 0, \text{ as } n \to \infty.$$

Therefore, letting $N_C = \sum_{j=0:\ n_j \le C}^{\ell-1} n_j$, the above bound, along with the fact $\mathcal{P}_0 = \{0\}$ yields $N_C/n \to 0$. With $C > 2\pi/(\theta_2 - \theta_1)$, the left side of (3.8) can be written as

$$\frac{1}{n} \sum_{j=0}^{\ell-1} \#\{s : \frac{2\pi(s + \Theta_j)}{n_j} \in [\theta_1, \theta_2], s = 1, 2, \ldots, n_j\}$$

$$\times \mathbb{I}\left(\left(\prod_{t \in \mathcal{P}_j} |\frac{1}{\sqrt{n}}\lambda_t| \right)^{1/n_j} \in (r - \epsilon, r + \epsilon) \right)$$

$$= \frac{n - N_C}{n} \frac{1}{n - N_C} \sum_{j=0,\ n_j > C}^{\ell-1} n_j \times n_j^{-1} \#\{s : \frac{s + \Theta_j}{n_j} \in \frac{1}{2\pi}[\theta_1, \theta_2], s = 1, \ldots, n_j\}$$

$$\times \mathbb{I}\left(\left(\prod_{t \in \mathcal{P}_j} |\frac{1}{\sqrt{n}}\lambda_t| \right)^{1/n_j} \in (r - \epsilon, r + \epsilon) \right) + O\left(\frac{N_C}{n}\right)$$

$$= \frac{1}{n - N_C} \sum_{j=0,\ n_j > C}^{\ell-1} n_j \left(\frac{(\theta_2 - \theta_1)}{2\pi} + O(C^{-1}) \right) \mathbb{I}\left(\left(\prod_{t \in \mathcal{P}_j} |\frac{\lambda_t}{\sqrt{n}}| \right)^{1/n_j} \in (r - \epsilon, r + \epsilon) \right)$$

$$+ O\left(\frac{N_C}{n}\right)$$

$$= \frac{1}{n - N_C} \sum_{j=0, n_j > C}^{\ell-1} n_j \times \frac{(\theta_2 - \theta_1)}{2\pi} \times \mathbb{I}\left(\left(\prod_{t \in \mathcal{P}_j} |\frac{1}{\sqrt{n}}\lambda_t| \right)^{1/n_j} \in (r - \epsilon, r + \epsilon) \right)$$

$$+ O(C^{-1}) + O\left(\frac{N_C}{n}\right)$$

$$= \frac{(\theta_2 - \theta_1)}{2\pi} + \frac{1}{n - N_C} \sum_{j=0,\, n_j > C}^{\ell - 1} n_j \times \mathbb{I}\left(\left(\prod_{t \in \mathcal{P}_j} |\frac{1}{\sqrt{n}} \lambda_t|\right)^{1/n_j} \notin (r - \epsilon, r + \epsilon)\right)$$

$$+ O(C^{-1}) + O\left(\frac{N_C}{n}\right). \tag{3.9}$$

To show that the second term in the above expression converges to zero in L_1, and hence in probability, it remains to prove,

$$P\left(\left(\prod_{t \in \mathcal{P}_j} |\frac{1}{\sqrt{n}} \lambda_t|\right)^{1/n_j} \notin (r - \epsilon, r + \epsilon)\right) \tag{3.10}$$

is uniformly small for all j such that $n_j > C$ and for all but finitely many n, provided we take C sufficiently large.

By Lemma 3.3.2, for each $1 \leq t < n$, $|\frac{1}{\sqrt{n}} \lambda_t|^2$ is an exponential random variable with mean one, and λ_t is independent of $\lambda_{t'}$ if $t' \neq n - t$ and $|\lambda_t| = |\lambda_{t'}|$ otherwise. Let $E, E_1, E_2, \ldots$ be i.i.d. exponential random variables with mean one. Observe that depending on whether or not $\mathcal{P}_j$ is conjugate to itself, (3.10) equals respectively,

$$P\left(\left(\prod_{t=1}^{n_j/2} E_t\right)^{1/n_j} \notin (r - \epsilon, r + \epsilon)\right) \quad \text{or} \quad P\left(\left(\prod_{t=1}^{n_j} \sqrt{E_t}\right)^{1/n_j} \notin (r - \epsilon, r + \epsilon)\right).$$

The theorem now follows by letting first $n \to \infty$ and then $C \to \infty$ in (3.9), and by observing that the *Strong Law of Large Numbers* (SLLN) implies that

$$\left(\prod_{t=1}^{C} \sqrt{E_t}\right)^{1/C} \to r = \exp\left(E\left[\log \sqrt{E}\right]\right) \quad \text{almost surely, as } C \to \infty. \quad \square$$

Let $\{E_i\}$ be i.i.d. $Exp(1)$, U_1 be uniformly distributed over $(2g)$-th roots of unity, U_2 be uniformly distributed over the unit circle where U_1, U_2 are independent of $\{E_i\}$. Then the following result can be established.

Theorem 3.3.3 (Bose et al. (2012b)). Suppose $\{x_l\}_{l \geq 0}$, satisfies Assumption 3.1.1. Fix $g \geq 1$ and let p_1 be the smallest prime divisor of g.

(a) Suppose $k^g = -1 + sn$ where $s = 1$ if $g = 1$, and $s = o(n^{p_1 - 1})$ if $g > 1$. Then $F_{n^{-1/2} A_{k,n}}$ converges weakly in probability to $U_1(\prod_{j=1}^{g} E_j)^{1/2g}$.

(b) Suppose $k^g = 1 + sn$ where $s = 0$ if $g = 1$, and $s = o(n^{p_1 - 1})$ if $g > 1$. Then $F_{n^{-1/2} A_{k,n}}$ converges weakly in probability to $U_2(\prod_{j=1}^{g} E_j)^{1/2g}$.

We skip the proof of the above theorem. However it will follow as a special case from the results with dependent inputs in the next chapter. Figures 3.3 and 3.4 provide simulations for $g = 2$ and $g = 5$ respectively.

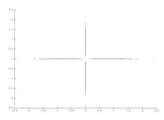 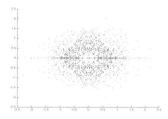

FIGURE 3.3
Eigenvalues of 20 realizations of $\frac{1}{\sqrt{n}}A_{k,n}$ with input i.i.d. $N(0,1)$ where (i) (left) $n = k^2+1$, $k = 10$ and (ii) (right) $n = k^2 - 1$, $k = 10$.

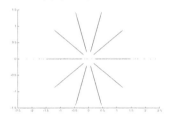 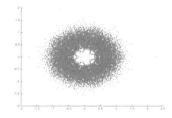

FIGURE 3.4
Eigenvalues of 20 realizations of $\frac{1}{\sqrt{n}}A_{k,n}$ with input i.i.d. $N(0,1)$ where (i) (left) $n = k^5+1$, $k = 4$ and (ii) (right) $n = k^5 - 1$, $k = 4$.

3.4   Exercises

1. Prove Remark 3.2.1.

2. Prove Theorem 3.2.2 using normal approximation.

3. Show that if $\{x_i\}$ satisfies Assumption 3.1.1, then the ESD of $\frac{1}{\sqrt{n}}PT_n$ converges weakly in L_2 to the standard normal distribution.

4. Prove Theorem 3.2.1 when the input sequence is complex and satisfies (3.3).

5. Show that $\sum_{\ell=0}^{n-1} \cos\left(\frac{2\pi j\ell}{n}\right) \sin\left(\frac{2\pi k\ell}{n}\right) = 0$ for all $0 \le j, k < n$ and use it to complete the proof of Lemma 3.3.2(a).

6. Simulate and generate the scatter plot of the eigenvalues of the k-circulant matrix for the cases:

   (a) $n = 1000$, $k = 2$,
   (b) $n = 1001$, $k = 2$,
   (c) $n = k^3 + 1$, $k = 10$,
   (d) $n = k^3 - 1$, $k = 10$.

4

LSD: dependent input

The class of stationary linear processes is an important class of dependent sequences. In this chapter the LSD results for i.i.d. inputs are extended to the situation where the input sequence is a stationary linear process. Under very modest conditions on the process, the LSDs for the circulant-type matrices exist and are interesting functions of the spectral density of the process.

4.1 Spectral density

The input sequence will be a *linear process* that satisfies the following assumptions.

Assumption 4.1.1. $\{x_n,\ n \geq 0\}$ is a two-sided linear process

$$x_n = \sum_{i=-\infty}^{\infty} a_i \varepsilon_{n-i}, \quad \text{where} \ \ a_n \in \mathbb{R} \ \ \text{and} \ \sum_{n \in \mathbb{Z}} |a_n| < \infty. \qquad (4.1)$$

Assumption 4.1.2. $\{\varepsilon_i,\ i \in \mathbb{Z}\}$ are i.i.d. random variables with mean zero, variance one, and $\mathrm{E}\,|\varepsilon_i|^{2+\delta} < \infty$ for some $\delta > 0$.

Recall the *autocovariance function* $\gamma(h) = \mathrm{Cov}(x_t, x_{t+h})$ which is finite under the above assumptions.

Definition 4.1.1. (i) Whenever the infinite sum below makes sense, the *spectral density* function of $\{x_n\}$ is defined as

$$\begin{aligned}
f(s) &= \frac{1}{2\pi} \sum_{k \in \mathbb{Z}} \gamma_k \exp(iks) \\
&= \frac{1}{2\pi} \Big[\gamma_0 + 2 \sum_{k \geq 1} \gamma_k \cos(ks) \Big], \quad \text{for} \ \ s \in [0, 2\pi].
\end{aligned}$$

(ii) The numbers $\omega_k = 2\pi k/n,\ k = 0, 1, \ldots, n-1$, are called the *Fourier frequencies*. The *periodogram* of $\{x_i\}$ is defined as

$$I_n(\omega_k) = \frac{1}{n} \Big| \sum_{t=0}^{n-1} x_t e^{-it\omega_k} \Big|^2, \quad k = 0, 1, \ldots, n-1. \qquad (4.2)$$

Under Assumptions 4.1.1 and 4.1.2, $\sum_{h=0}^{\infty} |\gamma(h)| < \infty$ and f is continuous. If f has some additional properties, the LSD of the circulant-type matrices exist. These LSDs are appropriate mixtures involving the spectral density f, the normal, the symmetrized Rayleigh, and other distributions. These results also reduce to the ones given in Chapter 3 when we specialize to i.i.d. input.

Throughout the chapter c and C denote generic constants.

4.2   Circulant

The following functions and matrices will be crucial to us. Let

$$\psi(e^{is}) = \sum_{j=-\infty}^{\infty} a_j e^{ijs}, \quad \psi_1(e^{is}) = \mathcal{R}[\psi(e^{is})], \quad \psi_2(e^{is}) = \mathcal{I}[\psi(e^{is})], \qquad (4.3)$$

where a_j's are as in (4.1). It is easy to see that

$$|\psi(e^{is})|^2 = [\psi_1(e^{is})]^2 + [\psi_2(e^{is})]^2 = 2\pi f(s).$$

Let

$$B(s) = \begin{pmatrix} \psi_1(e^{is}) & -\psi_2(e^{is}) \\ \psi_2(e^{is}) & \psi_1(e^{is}) \end{pmatrix}. \qquad (4.4)$$

Let N_1 and N_2 be i.i.d. standard normal variables. Define for $(x, y) \in \mathbb{R}^2$ and $s \in [0, 2\pi]$,

$$H_C(s, x, y) = \begin{cases} \mathrm{P}\left(B(s)(N_1, N_2)' \leq \sqrt{2}(x, y)'\right) & \text{if} \quad f(s) \neq 0, \\ \\ \mathbb{I}(x \geq 0, y \geq 0) & \text{if} \quad f(s) = 0. \end{cases}$$

Let

$$C_0 = \{t \in [0, 1] : f(2\pi t) = 0\} \quad \text{and} \quad \mathrm{Leb}(C_0) = \text{Lebesgue measure of } C_0.$$

Lemma 4.2.1. (a) For fixed x, y, H_C is a bounded continuous function in s.

(b) F_C defined below is a proper distribution function.

$$F_C(x, y) = \int_0^1 H_C(2\pi s, x, y) ds. \qquad (4.5)$$

(c) If $\mathrm{Leb}(C_0) = 0$ then F_C is continuous everywhere and equals

$$F_C(x, y) = \iint \mathbb{I}_{\{(v_1, v_2) \leq (x, y)\}} \left[\int_0^1 \frac{1}{2\pi^2 f(2\pi s)} e^{-\frac{v_1^2 + v_2^2}{2\pi f(2\pi s)}} ds \right] dv_1 dv_2. \qquad (4.6)$$

Further, F_C is bivariate normal if and only if f is constant a.e. (Lebesgue).

(d) If $\mathrm{Leb}(C_0) \neq 0$ then F_C is discontinuous *only* on $D_1 = \{(x, y) : xy = 0\}$.

Proof. The proof is easy. We omit the details and just show how the normality claim in (c) follows. If f is a constant function then it is easy to see that F_C is bivariate normal. Conversely, suppose F_C is bivariate normal. Let (X, Y) be a random vector with distribution function F_C. It is easy to see that

$$\mathrm{E}(X) = 0, \quad \mathrm{E}(X^2) = \pi \int_0^1 f(2\pi s)ds, \quad \text{and} \quad \mathrm{E}(X^4) = 3\pi^2 \int_0^1 f^2(2\pi s)ds.$$

On the other hand, since (X, Y) is bivariate normal, X is a normal random variable and hence

$$\mathrm{E}(X^4) = 3[\mathrm{E}(X^2)]^2 \Rightarrow 3\pi^2 \int_0^1 f^2(2\pi s)ds = 3\pi^2 \left(\int_0^1 f(2\pi s)ds \right)^2. \quad (4.7)$$

Now by an application of Cauchy-Schwarz inequality, (4.7) holds if and only if f is constant almost everywhere. $\qquad\square$

Theorem 4.2.2 (Bose et al. (2009)). Suppose Assumptions 4.1.1 and 4.1.2 hold. Then the ESD of $\frac{1}{\sqrt{n}}C_n$ converges in L_2 to $F_C(\cdot)$ given in (4.5)–(4.6).

Remark 4.2.1. If $\{x_i\}$ are i.i.d. with $\mathrm{E}\,|x_i|^{2+\delta} < \infty$, then $f(s) \equiv (2\pi)^{-1}$, and F_C is the bivariate normal distribution whose covariance matrix is diagonal with entries $1/2$ each. This agrees with Theorem 3.2.1.

Figure 4.1 provides a scatter plot of the eigenvalues of this matrix for $n = 1000$ when the input sequence is an MA(3) process.

Before going into the proof of Theorem 4.2.2, we observe a general fact which will be used in the proofs.

Lemma 4.2.3. Suppose $\{\lambda_{n,k}\}_{1 \le k \le n}$ is a triangular sequence of $\mathbb{R}^d$-valued random variables such that $\lambda_{n,k} = \eta_{n,k} + y_{n,k}$ for $1 \le k \le n$. Suppose F is a continuous distribution function such that

(i) $\lim_{n \to \infty} \frac{1}{n} \sum_{k=1}^{n} \mathrm{P}(\eta_{n,k} \le \tilde{x}) = F(\tilde{x})$ for $\tilde{x} \in \mathbb{R}^d$,

(ii) $\lim_{n \to \infty} \frac{1}{n^2} \sum_{k,l=1}^{n} \mathrm{P}(\eta_{n,k} \le \tilde{x}, \eta_{n,l} \le \tilde{y}) = F(\tilde{x})F(\tilde{y})$ for $\tilde{x}, \tilde{y} \in \mathbb{R}^d$, and

(iii) for any $\varepsilon > 0$, $\max_{1 \le k \le n} \mathrm{P}(|y_{n,k}| > \varepsilon) \to 0$ as $n \to \infty$.

Then

(a) $\lim_{n \to \infty} \frac{1}{n} \sum_{k=1}^{n} \mathrm{P}(\lambda_{n,k} \le \tilde{x}) = F(\tilde{x})$,

(b) $\lim_{n \to \infty} \frac{1}{n^2} \sum_{k,l=1}^{n} \mathrm{P}(\lambda_{n,k} \le \tilde{x}, \lambda_{n,l} \le \tilde{y}) = F(\tilde{x})F(\tilde{y})$.

Proof. Observe that

$$\frac{1}{n} \sum_{k=1}^{n} \mathrm{P}(\lambda_{n,k} \le \tilde{x})$$

$$= \frac{1}{n} \sum_{k=1}^{n} \mathrm{P}(\lambda_{n,k} \le \tilde{x}, |y_{n,k}| > \varepsilon) + \frac{1}{n} \sum_{k=1}^{n} \mathrm{P}(\lambda_{n,k} \le \tilde{x}, |y_{n,k}| \le \varepsilon). \quad (4.8)$$

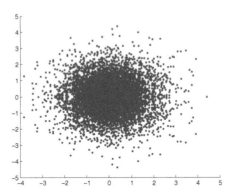

FIGURE 4.1
Eigenvalues of $\frac{1}{\sqrt{n}}C_n$ for $n = 1000$, 10 replications. Input sequence is the MA(3) process $x_t = \varepsilon_t + 0.5\varepsilon_{t-1} + 0.3\varepsilon_{t-2} + 0.2\varepsilon_{t-3}$ where $\{\varepsilon_t\}$ is i.i.d. $N(0, 1)$.

Now, by condition (iii),

$$\frac{1}{n}\sum_{k=1}^{n} P(\lambda_{n,k} \leq \tilde{x}, |y_{n,k}| > \varepsilon) \leq \frac{1}{n}\sum_{k=1}^{n} P(|y_{n,k}| > \varepsilon)$$

$$\leq \max_{1 \leq k \leq n} P(|y_{n,k}| > \varepsilon)$$

$$\to 0, \quad \text{as} \quad n \to \infty. \tag{4.9}$$

Also observe that

$$P(\lambda_{n,k} \leq \tilde{x}, |y_{n,k}| \leq \varepsilon) \leq P(\eta_{n,k} \leq \tilde{x} + \tilde{\varepsilon})$$

and

$$P(\lambda_{n,k} \leq \tilde{x}, |y_{n,k}| \leq \varepsilon) \geq P(\eta_{n,k} \leq \tilde{x} - \tilde{\varepsilon}) - P(|y_{n,k}| > \varepsilon),$$

where $\tilde{\varepsilon} = (\varepsilon,\ldots,\varepsilon) \in \mathbb{R}^d$. Now using conditions (i) and (iii), from the last two inequalities we get,

$$F(\tilde{x} - \tilde{\varepsilon}) \leq \lim_{n\to\infty} P(\lambda_{n,k} \leq \tilde{x}, |y_{n,k}| \leq \varepsilon) \leq F(\tilde{x} + \tilde{\varepsilon}). \tag{4.10}$$

Since $\varepsilon > 0$ can be chosen arbitrarily small, from (4.8), (4.9) and (4.10), we conclude that

$$\lim_{n\to\infty} \frac{1}{n}\sum_{k=1}^{n} P(\lambda_{n,k} \leq \tilde{x}) = F(\tilde{x}).$$

This completes the proof of (a). Proof of (b) is similar, so we skip it. $\qquad\square$

We now proceed to prove Theorem 4.2.2. The normal approximation Lemma 3.1.1 and the result from Fan and Yao (2003) quoted below allow us to approximate the eigenvalues by appropriate partial sums of independent random variables.

Theorem 4.2.4 (Fan and Yao (2003)). Suppose that $\{x_i\}$ is a sample from a stationary process defined in (4.1), where $\{\varepsilon_t\}$ are i.i.d. random variables with mean zero and variance 1. For $k = 1, 2, \ldots, \lfloor \frac{n-1}{2} \rfloor$, define

$$\xi_{2k-1} = \frac{1}{\sqrt{n}} \sum_{t=0}^{n-1} \varepsilon_t \cos(\omega_k t), \quad \xi_{2k} = \frac{1}{\sqrt{n}} \sum_{t=0}^{n-1} \varepsilon_t \sin(\omega_k t), \quad \text{where } \omega_k = \frac{2\pi k}{n}.$$

(a) As $n \to \infty$, $\{\xi_k; k = 1, 2, \ldots, n\}$ is a sequence of asymptotically i.i.d. $N(0, 1/2)$ in the sense that for any integer $r \geq 1$, any $c_1, c_2, \ldots, c_r \in \mathbb{R}$, and any $k_r > \cdots > k_1 \geq 1$,

$$\sum_{j=1}^{r} c_j \xi_{k_j} \xrightarrow{D} N\left(0, \frac{1}{2} \sum_{j=1}^{r} c_j^2\right).$$

(b) For $k = 1, 2, \ldots, n$,

$$I_n(\omega_k) = \frac{1}{n} \left| \sum_{t=0}^{n-1} x_t e^{-it\omega_k} \right|^2 = L_n(\omega_k) + R_n(\omega_k),$$

where

$$L_n(\omega_k) = 2\pi g(\omega_k)(\xi_{2k-1}^2 + \xi_{2k}^2), \tag{4.11}$$

$g(\cdot)$ is the spectral density of $\{x_t\}$ and

$$\lim_{n \to \infty} \max_{1 \leq k \leq n} E|R_n(\omega_k)| = 0.$$

The next lemma is similar to Theorem 4.2.4(b). We provide a proof for the sake of completeness.

Lemma 4.2.5. Suppose Assumption 4.1.1 holds and $\{\varepsilon_t\}$ are i.i.d. random variables with mean 0 and variance 1. For $k = 1, 2, \ldots, n$, write

$$\frac{1}{\sqrt{n}} \sum_{l=0}^{n-1} x_l e^{i\omega_k l} = \psi(e^{i\omega_k})[\xi_{2k-1} + i\xi_{2k}] + Y_n(\omega_k),$$

where ξ_{2k-1} and ξ_{2k} are as in Theorem 4.2.4. Then, as $n \to \infty$, we have $\max_{1 \leq k < n} E|Y_n(\omega_k)| \to 0$.

Proof. Let

$$U_{nj} = \sum_{t=-j}^{n-1-j} \varepsilon_t e^{i\omega_k t} - \sum_{t=0}^{n-1} \varepsilon_t e^{i\omega_k t}, \quad Y_n(\omega_k) = \frac{1}{\sqrt{n}} \sum_{j=-\infty}^{\infty} a_j e^{i\omega_k j} U_{nj}.$$

Now observe that

$$\frac{1}{\sqrt{n}} \sum_{t=0}^{n-1} x_t e^{i\omega_k t} = \frac{1}{\sqrt{n}} \sum_{j=-\infty}^{\infty} a_j e^{i\omega_k j} \sum_{t=0}^{n-1} \varepsilon_{t-j} e^{i\omega_k(t-j)}$$

$$= \frac{1}{\sqrt{n}} \sum_{j=-\infty}^{\infty} a_j e^{i\omega_k j} \left(\sum_{t=0}^{n-1} \varepsilon_t e^{i\omega_k t} + U_{nj} \right)$$

$$= \psi(e^{i\omega_k})[\xi_{2k-1} + i\xi_{2k}] + Y_n(\omega_k).$$

Note that if $|j| < n$, U_{nj} is a sum of $2|j|$ independent random variables, whereas if $|j| \geq n$, U_{nj} is a sum of $2n$ independent random variables. Thus

$$E|U_{nj}|^2 \leq 2 \min(|j|, n).$$

Therefore, for any fixed positive integer l and $n > l$,

$$E|Y_n(\omega_k)| \leq \frac{1}{\sqrt{n}} \sum_{j=-\infty}^{\infty} |a_j| (E U_{nj}^2)^{1/2}$$

$$\leq \sqrt{\frac{2}{n}} \sum_{j=-\infty}^{\infty} |a_j| \{\min(|j|, n)\}^{1/2}$$

$$\leq \sqrt{2} \left(\frac{1}{\sqrt{n}} \sum_{|j| \leq l} |a_j| |j|^{1/2} + \sum_{|j| > l} |a_j| \right).$$

The last expression is free of k. First choose l large enough so that the second sum is small. Now as $n \to \infty$, the first sum goes to zero. Hence

$$\max_{1 \leq k \leq n} E|Y_n(\omega_k)| \to 0.$$

$\square$

Proof of Theorem 4.2.2. As pointed out in (3.2) of Chapter 3, to prove that the ESD F_n converges to an F in L_2, it is enough to show that at all continuity points (x, y) of F,

$$E[F_n(x, y)] \to F(x, y) \text{ and } \mathrm{Var}[F_n(x, y)] \to 0. \tag{4.12}$$

This is what we show here and in every proof later on. First assume that $\mathrm{Leb}(C_0) = 0$. Recall the eigenvalues of C_n from Section 1.1 of Chapter 1. As before we may ignore the eigenvalue λ_n and also $\lambda_{n/2}$ (when n is even). So for $x, y \in \mathbb{R}$,

$$\mathrm{E}[F_n(x,y)] \sim \frac{1}{n} \sum_{k=1, k \neq n/2}^{n-1} \mathrm{P}(b_k \leq x, c_k \leq y),$$

where

$$b_k = \frac{1}{\sqrt{n}} \sum_{j=0}^{n-1} x_j \cos(\omega_k j), \quad c_k = \frac{1}{\sqrt{n}} \sum_{j=0}^{n-1} x_j \sin(\omega_k j), \quad \omega_k = \frac{2\pi k}{n}.$$

Define for $k = 1, 2, \ldots, n$,

$$\eta_k = (\xi_{2k-1}, \xi_{2k})', \quad Y_{1n}(\omega_k) = \mathcal{R}[Y_n(\omega_k)], \quad Y_{2n}(\omega_k) = \mathcal{I}[Y_n(\omega_k)],$$

where $Y_n(\omega_k)$ are the same as defined in Lemma 4.2.5. Then

$$(b_k, c_k)' = B(\omega_k)\eta_k + (Y_{1n}(\omega_k), Y_{2n}(\omega_k))'.$$

Now in view of Lemmata 4.2.3 and 4.2.5, to conclude that $\mathrm{E}[F_n(x,y)] \to F_C(x,y)$ it is sufficient to show that

$$\frac{1}{n} \sum_{k=1, k \neq n/2}^{n-1} \mathrm{P}\left[B(\omega_k)\eta_k \leq (x,y)'\right] \to F_C(x,y). \tag{4.13}$$

For this, define for $1 \leq k \leq n - 1$, (except for $k = n/2$) and $0 \leq l \leq n - 1$,

$$X_{l,k} = \left(\sqrt{2}\varepsilon_l \cos(\omega_k l), \ \sqrt{2}\varepsilon_l \sin(\omega_k l)\right)'.$$

Note that

$$\mathrm{E}(X_{l,k}) = 0, \tag{4.14}$$

$$\frac{1}{n} \sum_{l=0}^{n-1} \mathrm{Cov}(X_{l,k}) = I, \text{ and} \tag{4.15}$$

$$\sup_n \sup_{1 \leq k \leq n} \left[\frac{1}{n} \sum_{l=0}^{n-1} \mathrm{E} \| X_{l,k} \|^{(2+\delta)}\right] \leq C < \infty. \tag{4.16}$$

For $k \neq n/2$,

$$\{B(\omega_k)\eta_k \leq (x,y)'\} = \left\{B(\omega_k)\left(\frac{1}{\sqrt{n}} \sum_{l=0}^{n-1} X_{l,k}\right) \leq (\sqrt{2}x, \sqrt{2}y)'\right\}.$$

Since

$$\{(r,s) : B(\omega_k)(r,s)' \leq (\sqrt{2}x, \sqrt{2}y)'\}$$

is a *convex* set in $\mathbb{R}^2$, and $\{X_{l,k}, \; l = 0, 1, \ldots, (n-1)\}$ satisfies (4.14)–(4.16), we can apply Lemma 3.1.1(a) for $k \neq n/2$ to get, with $z = (\sqrt{2}x, \sqrt{2}y)'$,

$$\left| \mathrm{P}\left(B(\omega_k)(\frac{1}{\sqrt{n}} \sum_{l=0}^{n-1} X_{l,k}) \leq z\right) - \mathrm{P}\left(B(\omega_k)(N_1, \; N_2)' \leq z\right) \right|$$

$$\leq Cn^{-\delta/2}[\frac{1}{n} \sum_{l=0}^{n-1} \mathrm{E} \, \|X_{lk}\|^{(2+\delta)}]$$

$$\leq Cn^{-\delta/2} \to 0, \text{ as } n \to \infty.$$

Since by Lemma 4.2.1(a), H_C is bounded continuous for every fixed (x, y),

$$\lim_{n\to\infty} \frac{1}{n} \sum_{k=1, k\neq n/2}^{n-1} \mathrm{P}\left(B(\omega_k)\eta_k \leq (x, y)'\right) = \lim_{n\to\infty} \frac{1}{n} \sum_{k=1, k\neq n/2}^{n-1} H_C(\frac{2\pi k}{n}, x, y)$$

$$= \int_0^1 H_C(2\pi s, x, y)ds$$

$$= F_C(x, y).$$

Hence, by (4.13),

$$\mathrm{E}[F_n(x, y)] \to \int_0^1 H_C(2\pi s, x, y)ds = F_C(x, y). \tag{4.17}$$

To show $\mathrm{Var}[F_n(x, y)] \to 0$, it is enough to show that

$$\frac{1}{n^2} \sum_{k\neq k'=1}^{n} \mathrm{Cov}(J_k, J_{k'}) = \frac{1}{n^2} \sum_{k\neq k'=1}^{n} [\mathrm{E}(J_k, J_{k'}) - \mathrm{E}(J_k) \mathrm{E}(J_{k'})]$$

$$\to 0, \tag{4.18}$$

where for $1 \leq k \leq n$, J_k is the indicator that $\{b_k \leq x, c_k \leq y\}$. As $n \to \infty$,

$$\frac{1}{n^2} \sum_{k\neq k'=1}^{n} \mathrm{E}(J_k) \mathrm{E}(J_{k'}) = [\frac{1}{n} \sum_{k=1}^{n} \mathrm{E}(J_k)]^2 - \frac{1}{n^2} \sum_{k=1}^{n} [\mathrm{E}(J_k)]^2$$

$$\to \left[\int_0^1 H_C(2\pi s, x, y)ds\right]^2.$$

So, to show (4.18), it is enough to show that as $n \to \infty$,

$$\frac{1}{n^2} \sum_{k\neq k'=1}^{n} \mathrm{E}(J_k, J_{k'}) \to \left[\int_0^1 H_C(2\pi s, x, y)ds\right]^2.$$

Similar to the proof of (4.17), instead of vectors with two coordinates, define appropriate vectors with four coordinates and proceed as above to verify this. We omit the details. This completes the proof for the case $\mathrm{Leb}(C_0) = 0$.

If $\mathrm{Leb}(C_0) \neq 0$, we have to show (4.12) on D_1^c (of Lemma 4.2.1). All the above steps in the proof remain valid for $(x, y) \in D_1^c$. Hence, if $\mathrm{Leb}(C_0) \neq 0$, we have our required LSD. This completes the proof of Theorem 4.2.2. $\qquad\square$

4.3   Reverse circulant

Let G be the standard exponential distribution function,

$$G(x) = 1 - e^{-x}, \ x > 0.$$

Define $H_R(s, x)$ on $[0, 2\pi] \times \mathbb{R}$ as

$$H_R(s, x) = \begin{cases} G\left(\frac{x^2}{2\pi f(s)}\right) & \text{if} \quad f(s) \neq 0 \\ 1 & \text{if} \quad f(s) = 0. \end{cases}$$

Let

$$S_0 = \{t \in [0, 1/2] : f(2\pi t) = 0\}.$$

The following lemma is analogous to Lemma 4.2.1. We omit its proof.

Lemma 4.3.1. (a) For fixed x, $H_R(s, x)$ is bounded continuous on $[0, 2\pi]$.

(b) F_R defined below is a valid distribution function.

$$F_R(x) = \begin{cases} \frac{1}{2} + \int_0^{1/2} H_R(2\pi t, x)dt & \text{if} \quad x > 0 \\ \frac{1}{2} - \int_0^{1/2} H_R(2\pi t, x)dt & \text{if} \quad x \leq 0. \end{cases} \tag{4.19}$$

(c) If $\mathrm{Leb}(S_0) = 0$ then F_R is continuous everywhere and equals

$$F_R(x) = \begin{cases} 1 - \int_0^{1/2} e^{-\frac{x^2}{2\pi f(2\pi t)}} dt & \text{if} \quad x > 0 \\ \int_0^{1/2} e^{-\frac{x^2}{2\pi f(2\pi t)}} dt & \text{if} \quad x \leq 0. \end{cases} \tag{4.20}$$

Further, F_R equals $\mathcal{L}_R$ given in (2.17) $\Leftrightarrow f$ is constant almost everywhere (Lebesgue).

(d) If $\mathrm{Leb}(S_0) \neq 0$ then F_R is discontinuous *only* at $x = 0$.

Theorem 4.3.2 (Bose et al. (2009)). If Assumptions 4.1.1 and 4.1.2 hold then the ESD of $\frac{1}{\sqrt{n}} RC_n$ converges in L_2 to F_R given in (4.19)–(4.20).

Remark 4.3.1. If $\{x_i\}$ are i.i.d., with $\mathrm{E}(|x_i|^{2+\delta}) < \infty$, then $f(s) \equiv 1/2\pi$ and the LSD $F_R(\cdot)$ agrees with $\mathcal{L}_R$ given in (2.17) of Section 2.3, Chapter 2.

Figure 4.2 shows the histogram of the simulated eigenvalues of $\frac{1}{\sqrt{n}} RC_n$ when the input sequence is an MA(1) process.

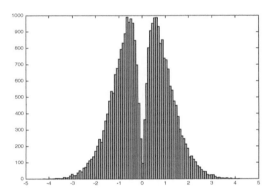

FIGURE 4.2

Histogram of the eigenvalues of $\frac{1}{\sqrt{n}}RC_n$ for $n = 1000$, 30 replications. Input sequence is the MA(1) process $x_t = \varepsilon_t + 0.5\varepsilon_{t-1}$ where $\{\varepsilon_t\}$ is i.i.d. $N(0, 1)$.

Proof of Theorem 4.3.2. As before, we give the proof only when $\text{Leb}(S_0) = 0$. The structure of the eigenvalues $\{\lambda_k\}$ of RC_n (see Section 1.3 of Chapter 1) implies that the LSD is going to be symmetric about 0. As before we may also ignore the two eigenvalues λ_0 and $\lambda_{n/2}$. Hence for $x > 0$, with $\omega_k = 2\pi k/n$,

$$E[F_n(x)] \sim 1/2 + \frac{1}{n}\sum_{k=1}^{\lfloor\frac{n-1}{2}\rfloor} P(\frac{1}{n}\lambda_k^2 \leq x^2)$$

$$= 1/2 + \frac{1}{n}\sum_{k=1}^{\lfloor\frac{n-1}{2}\rfloor} P(I_n(\omega_k) \leq x^2),$$

where $I_n(\omega_k)$ is as in (4.2). Along the same lines as in the proof of Theorem 4.2.2, using Lemma 4.2.3 and Theorem 4.2.4(b), it is sufficient to show that

$$\frac{1}{n}\sum_{k=1}^{\lfloor\frac{n-1}{2}\rfloor} P\left(L_n(\omega_k) \leq x^2\right) \to \int_0^{1/2} H_R(2\pi t, x)dt,$$

where

$$L_n(\omega_k) = 2\pi f(\omega_k)(\xi_{2k-1}^2 + \xi_{2k}^2), \text{ as in (4.11)}.$$

For $1 \leq k \leq \lfloor\frac{n-1}{2}\rfloor$ and $0 \leq l \leq n-1$, let

$$X_{l,k} = \left(\sqrt{2}\varepsilon_l\cos(l\omega_k), \ \sqrt{2}\varepsilon_l\sin(l\omega_k)\right)',$$

$$A_{kn} = \left\{(r_1, r_2) : \pi f(\omega_k)(r_1^2 + r_2^2) \leq x^2\right\}.$$

Note that $\{X_{l,k}\}$ satisfies (4.14)–(4.16) and

$$\{L_n(\omega_k) \leq x^2\} = \{\frac{1}{\sqrt{n}}\sum_{l=0}^{n-1} X_{l,k} \in A_{kn}\}.$$

Since A_{kn} is a *convex* set in $\mathbb{R}^2$, we can apply Lemma 3.1.1(a) to get

$$\frac{1}{n} \sum_{k=1}^{\lfloor \frac{n-1}{2} \rfloor} |P(L_n(\omega_k) \le x^2) - \Phi_{0,I}(A_{kn})| \le Cn^{-\delta/2} \to 0.$$

But

$$\frac{1}{n} \sum_{k=1}^{\lfloor \frac{n-1}{2} \rfloor} \Phi_{0,I}(A_{kn}) \;=\; \frac{1}{n} \sum_{k=1}^{\lfloor \frac{n-1}{2} \rfloor} H_R(\frac{2\pi k}{n}, x) \to \int_0^{1/2} H_R(2\pi t, x)dt.$$

Hence for $x \ge 0$,

$$E[F_n(x)] \to \frac{1}{2} + \int_0^{1/2} H_R(2\pi t, x)dt = F_R(x).$$

The rest of the argument is same as in the proof of Theorem 4.2.2. We leave it as an exercise. $\qquad\square$

4.4 Symmetric circulant

For $x \in \mathbb{R}$ and $s \in [0, \pi]$, define

$$H_S(s, x) = \begin{cases} P\left(\sqrt{2\pi f(s)}N(0,1) \le x\right) & \text{if } f(s) \ne 0, \\ \mathbb{I}(x \ge 0) & \text{if } f(s) = 0. \end{cases} \tag{4.21}$$

Let

$$S_0 = \{t \in [0, 1/2] : f(2\pi t) = 0\}.$$

The following lemma is analogous to Lemma 4.3.1. We omit its proof.

Lemma 4.4.1. (a) For fixed x, H_S is a bounded continuous function in s and

$$H_S(s, x) + H_S(s, -x) = 1.$$

(b) F_S given below is a proper distribution function and $F_S(x) + F_S(-x) = 1$.

$$F_S(x) = 2 \int_0^{1/2} H_S(2\pi s, x)ds. \tag{4.22}$$

(c) If $\text{Leb}(S_0) = 0$ then F_S is continuous everywhere and equals

$$F_S(x) = \int_{-\infty}^x \left[\int_0^{1/2} \frac{1}{\pi\sqrt{f(2\pi s)}} e^{-\frac{t^2}{4\pi f(2\pi s)}} ds \right] dt. \tag{4.23}$$

Further, F_S is normal if and only if f is constant almost everywhere (Lebesgue).

(d) If $\text{Leb}(S_0) \ne 0$ then F_S is discontinuous *only* at $x = 0$.

Theorem 4.4.2 (Bose et al. (2009)). Suppose Assumptions 4.1.1 and 4.1.2 hold and

$$\lim_{n\to\infty} \frac{1}{n^2} \sum_{k=1}^{\lfloor np/2 \rfloor} [f(\frac{2\pi k}{n})]^{-3/2} \to 0 \ \text{ for all } \ 0 < p < 1. \tag{4.24}$$

Then the ESD of $\frac{1}{\sqrt{n}} SC_n$ converges in L_2 to F_S given in (4.22)–(4.23).

Figure 4.3 shows the histogram of the simulated eigenvalues of $\frac{1}{\sqrt{n}} SC_n$ when the input sequence is an MA(1) process.

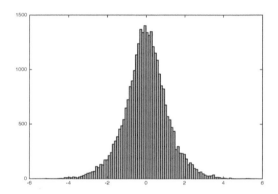

FIGURE 4.3
Histogram of the eigenvalues of $\frac{1}{\sqrt{n}} SC_n$ for $n = 1000$, 30 replications. Input sequence is the MA(1) process $x_t = \varepsilon_t + 0.5\varepsilon_{t-1}$ where $\{\varepsilon_t\}$ is i.i.d. $N(0, 1)$.

Remark 4.4.1. If $\inf_{s\in[0,2\pi]} f(s) > 0$, then (4.24) holds. It is not clear whether the LSD result will be true if (4.24) does not hold. If $f \equiv \frac{1}{2\pi}$ then F_S is the standard normal distribution function. This agrees with the conclusion of Theorem 3.2.2.

We give the proof only for odd $n = 2m + 1$. The proof in the even case is similar. First recall the eigenvalues of SC_n from Section 1.2 of Chapter 1. The approximation Lemma 4.2.5 now takes the following form. Its proof is left as an exercise.

Lemma 4.4.3. Suppose Assumption 4.1.1 holds and $\{\varepsilon_t\}$ are i.i.d. random variables with mean 0 and variance 1. For $n = 2m + 1$ and $1 \leq k \leq m$, write

$$\sum_{t=1}^{m} x_t \cos\frac{2\pi kt}{n} = \psi_1(e^{i\omega_k}) \sum_{t=1}^{m} \varepsilon_t \cos\frac{2\pi kt}{n} - \psi_2(e^{i\omega_k}) \sum_{t=1}^{m} \sin\frac{2\pi kt}{n} + \sqrt{n}Y_{n,k},$$

where $\psi_1(e^{i\omega_k})$, $\psi_2(e^{i\omega_k})$ are as in (4.3). Then $\max_{1\leq k\leq m} \mathrm{E}(|Y_{n,k}|) \to 0$.

Proof of Theorem 4.4.2. All eigenvalues $\{\lambda_k,\ 0 \le k \le n-1\}$ are real in this case. As before, we provide the detailed proof only when $\mathrm{Leb}(S_0) = 0$. As before the eigenvalue λ_0 can be ignored. Further, the term $\frac{x_0}{\sqrt{n}}$ can be ignored from the expression of the eigenvalues $\{\lambda_k\}$. So for $x \in \mathbb{R}$,

$$\mathrm{E}[F_n(x)] \sim \frac{2}{n}\sum_{k=1}^{m} \mathrm{P}(\frac{1}{\sqrt{n}}\lambda_k \le x) \sim \frac{2}{n}\sum_{k=1}^{m} \mathrm{P}\left(\frac{1}{\sqrt{n}}\sum_{t=1}^{m} 2x_t \cos\frac{2\pi kt}{n} \le x\right).$$

To show that $\mathrm{E}[F_n(x)] \to F_S(x)$, following the argument given in the circulant case, and using Lemmata 4.2.3 and 4.4.3, it is sufficient to show that,

$$\frac{2}{n}\sum_{k=1}^{m} \mathrm{P}\left[\psi_1(e^{i\omega_k})\frac{2}{n}\sum_{t=1}^{m} \varepsilon_t \cos\frac{2\pi kt}{n} - \psi_2(e^{i\omega_k})\frac{2}{\sqrt{n}}\sum_{t=1}^{m}\sin\frac{2\pi kt}{n} \le x\right]$$

$$= \frac{2}{n}\sum_{t=1}^{m}\mathrm{P}\left\{m^{-1/2}\sum_{l=1}^{m} X_{l,k} \in C_k\right\} \to F_S(x),$$

where

$$X_{l,k} = \left(2\sigma_n^{-1}\varepsilon_l\cos\frac{2\pi kl}{n},\ 2\delta_n^{-1}\varepsilon_l\sin\frac{2\pi kl}{n}\right),\ \sigma_n^2 = 2 - 1/m,\ \delta_n^2 = 2 + 1/m,$$

$$C_k = \{(u,v): \sigma_n\psi_1(e^{i\omega_k})u + \delta_n\psi_2(e^{i\omega_k})v \le \sqrt{n/m}x\}.$$

Let

$$V_k = \begin{pmatrix} 1 & -\frac{1}{\sqrt{4m^2-1}}\tan\frac{k\pi}{2m+1} \\ -\frac{1}{\sqrt{4m^2-1}}\tan\frac{k\pi}{2m+1} & 1 \end{pmatrix}.$$

Note that

$$\mathrm{E}(X_{l,k}) = 0,\quad \frac{1}{m}\sum_{l=1}^{m}\mathrm{Cov}(X_{l,k}) = V_k,\quad \sup_{1 \le k \le m}\frac{1}{m}\sum_{l=1}^{m}\mathrm{E}\,\|X_{l,k}\|^{2+\delta} \le C < \infty.$$

$$(4.25)$$

Let α_k be the minimum eigenvalue of V_k. Then $\alpha_k \ge \alpha_m$ for $1 \le k \le m$, and

$$\alpha_m = 1 - \frac{1}{\sqrt{4m^2-1}}\tan\frac{m\pi}{2m+1} \approx 1 - \frac{2m+1}{m\pi} \approx 1 - \frac{2}{\pi} = \alpha,\ \text{say}.$$

Since $\{X_{l,k}\}$ satisfies (4.25) and C_k is a *convex* set in $\mathbb{R}^2$, we can apply Lemma 3.1.1(a) for $k = 1, 2, \ldots, m$, to get

$$\left|\frac{2}{n}\sum_{k=1}^{m}\left[\mathrm{P}\left\{m^{-1/2}\sum_{l=1}^{m} X_{l,k} \in C_k\right\} - \Phi_{0,V_k}(C_k)\right]\right| \le Cm^{-\delta/2}\frac{2}{n}\sum_{k=1}^{m}\alpha_k^{-3/2}$$

$$\le Cm^{-\delta/2}\alpha^{-3/2} \to 0,$$

where Φ_{0,V_k} is a bivariate normal distribution with mean zero and covariance matrix V_k. Note that for large m, $\sigma_n^2 \approx 2$ and $\delta_n^2 \approx 2$. Hence

$$C_k' = \{(u,v): \psi_1(e^{i\omega_k})u + \psi_2(e^{i\omega_k})v \le \sqrt{x}\}$$

serves as a good approximation to C_k and we get

$$\frac{2}{n} \sum_{k=1}^{m} \Phi_{0,V_k}(C_k) \sim \frac{2}{n} \sum_{k=1}^{m} \Phi_{0,V_k}(C'_k) = \frac{2}{n} \sum_{k=1}^{m} \mathrm{P}(\mu_k N(0,1) \le x),$$

where

$$\mu_k^2 = \psi_1(e^{i\omega_k})^2 + \psi_2(e^{i\omega_k})^2 + 2\psi_1(e^{i\omega_k})\psi_2(e^{i\omega_k}) \frac{1}{\sqrt{4m^2-1}} \tan \frac{k\pi}{2m+1}.$$

Define

$$\nu_k^2 = \psi_1(e^{i\omega_k})^2 + \psi_2(e^{i\omega_k})^2.$$

Now we show that

$$\lim_{n\to\infty} \left| \frac{2}{n} \sum_{k=1}^{m} \left[\mathrm{P}(\mu_k N(0,1) \le x) - \mathbb{P}(\nu_k N(0,1) \le x) \right] \right| = 0. \tag{4.26}$$

Let $0 < p < 1$. Now as $n \to \infty$, using Assumption (4.24), the above expression, summed only for $1 \le k \le \lfloor mp \rfloor$, equals

$$\frac{2}{n} \sum_{k=1}^{\lfloor mp \rfloor} \left| \int_{x/\nu_k}^{x/\mu_k} \frac{1}{\sqrt{2\pi}} e^{-\frac{t^2}{2}} dt \right| \le \frac{2|x|}{n} \sum_{k=1}^{\lfloor mp \rfloor} \left| \frac{\mu_k^2 - \nu_k^2}{\mu_k \nu_k (\mu_k + \nu_k)} \right|$$

$$\le \frac{2|x| \tan \frac{p\pi}{2}}{m^2} \sum_{k=1}^{\lfloor mp \rfloor} \frac{1}{\nu_k^3 \alpha (1+\alpha)} \to 0.$$

On the other hand, for every n,

$$\frac{2}{n} \left| \sum_{\lfloor mp \rfloor + 1}^{m} \left[\mathrm{P}(\mu_k N(0,1) \le x) - \mathrm{P}(\nu_k N(0,1) \le x) \right] \right| \le 4(1-p).$$

Therefore, by letting first $n \to \infty$ and then $p \to 1$, (4.26) holds. Hence

$$\lim_{n\to\infty} \frac{2}{n} \sum_{k=1}^{m} \mathrm{P}\left(\nu_k N(0,1) \le x\right) = \lim_{n\to\infty} \frac{2}{n} \sum_{k=1}^{m} \mathrm{P}\left(\sqrt{2\pi f(2\pi k/n)} N(0,1) \le x\right)$$

$$\to 2 \int_0^{1/2} H_S(2\pi s, x) ds.$$

The rest of the argument in the proof is the same as in the proof of Theorem 4.2.2. $\qquad\square$

4.5 k-circulant

First recall the eigenvalues of the k-circulant matrix $A_{k,n}$ and related notation from Section 1.4 of Chapter 1. For any positive integers k, n, let $p_1 < p_2 <$

$\cdots < p_c$ be all their common prime factors and write

$$n = n' \prod_{q=1}^{c} p_q^{\beta_q} \quad \text{and} \quad k = k' \prod_{q=1}^{c} p_q^{\alpha_q}.$$

Here α_q, $\beta_q \geq 1$ and n', k', p_q are pairwise relatively prime. Then the characteristic polynomial of $A_{k,n}$ is given by

$$\chi(A_{k,n}) = \lambda^{n-n'} \prod_{j=0}^{\ell-1} (\lambda^{n_j} - y_j), \qquad (4.27)$$

where y_j, n_j are as in (1.8) and (1.9). Also recall from Section 1.4 of Chapter 1 that for $0 \leq x < n'$,

$$S(x) = \{xk^b \bmod n' : b \geq 0\}, \quad g_x = \#S(x), \quad \upsilon_{k,n'} = \#\{x \in \mathbb{Z}_{n'} : g_x < g_1\}.$$

As mentioned before, there is no general result on the LSD for all possible pairs (k, n). We give LSD results for two subclasses of the k-circulant matrix, where (k, n) satisfies either $n = k^g + 1$ or $n = k^g - 1$ and $g \geq 2$. Note that in both the cases $\gcd(n, k) = 1$, and hence $n' = n$ in (4.27).

We first state two lemmata, due to Bose et al. (2012b), which shall be used in the proof of the forthcoming theorems, and also in Chapters 6 and 8.

Lemma 4.5.1. Let b and c be two fixed positive integers. Then for any integer $k \geq 2$, the following inequality holds in each of the four cases:

$$\gcd(k^b \pm 1, k^c \pm 1) \leq k^{\gcd(b,c)} + 1.$$

Proof. The assertion trivially follows if one of b and c divides the other. So, we assume, without loss, that $b < c$ and b does not divide c. Since $k^c \pm 1 = k^{c-b}(k^b + 1) + (-k^{c-b} \pm 1)$, we can write

$$\gcd(k^b + 1, k^c \pm 1) = \gcd(k^b + 1, k^{c-b} \mp 1).$$

Similarly,

$$\gcd(k^b - 1, k^c \pm 1) = \gcd(k^b - 1, k^{c-b} \pm 1).$$

Moreover, if we write $c_1 = c - \lfloor c/b \rfloor b$, then by repeating the above step $\lfloor c/b \rfloor$ times, we can see that $\gcd(k^b \pm 1, k^c \pm 1)$ is equal to one of $\gcd(k^b \pm 1, k^{c_1} \pm 1)$. Now if c_1 divides b, then $\gcd(b, c) = c_1$ and we are done. Otherwise, we can now repeat the whole argument with $b = c_1$ and $c = b$ to deduce that $\gcd(k^b \pm 1, k^{c_1} \pm 1)$ is one of $\gcd(k^{b_1} \pm 1, k^{c_1} \pm 1)$ where $b_1 = b - \lfloor b/c_1 \rfloor c_1$. We continue in a similar fashion, reducing each time one of the two exponents of k in the gcd and the lemma follows once we recall Euclid's recursive algorithm for computing the gcd of two numbers. $\square$

Lemma 4.5.2. (a) Fix $g \geq 1$. Suppose $k^g = -1 + sn$, $n \to \infty$ with $s = 1$ if $g = 1$ and $s = o(n^{p_1-1})$ if $g > 1$ where p_1 is the smallest prime divisor of g. Then $g_1 = 2g$ for all but finitely many n, and $\frac{v_{k,n}}{n} \to 0$.

(b) Suppose $k^g = 1 + sn$, $g \geq 1$ fixed, $n \to \infty$ with $s = 0$ if $g = 1$ and $s = o(n^{p_1-1})$ where p_1 is the smallest prime divisor of g. Then $g_1 = g$ for all but finitely many n, and $\frac{v_{k,n}}{n} \to 0$.

Proof. (a) First note that $\gcd(n,k) = 1$ and therefore $n' = n$. When $g = 1$, it is easy to check that $g_1 = 2$ and by Lemma 1.4.4(a), $v_{k,n} \leq 2$.

Now assume $g > 1$. Since $k^{2g} = (sn - 1)^2 = 1 \pmod{n}$, g_1 divides $2g$. Observe that $g_1 \neq g = 2g/2$ because $k^g = -1 \pmod{n}$.

If $g_1 = 2g/b$, where b divides g and $b \geq 3$, then by Lemma 4.5.1,

$$\gcd(k^{g_1} - 1, n) = \gcd\left(k^{2g/b} - 1, (k^g + 1)/s\right) \leq \gcd\left(k^{2g/b} - 1, k^g + 1\right)$$
$$\leq k^{\gcd(2g/b,\, g)} + 1.$$

Note that since $\gcd(2g/b, g)$ divides g and $\gcd(2g/b, g) \leq 2g/b < g$, we have $\gcd(2g/b, g) \leq g/p_1$. Consequently,

$$\gcd(k^{2g/b} - 1, n) \leq k^{g/p_1} + 1 \leq (sn - 1)^{1/p_1} + 1 = o(n), \qquad (4.28)$$

which is a contradiction to the fact that $k^{g_1} = 1 \pmod{n}$. This implies that $\gcd(k^{g_1} - 1, n) = n$. Hence $g_1 = 2g$. Now by Lemma 1.4.4(b) it is enough to show that for any fixed $b \geq 3$ so that b divides g_1,

$$\gcd(k^{g_1/b} - 1, n)/n = o(1) \quad \text{as} \quad n \to \infty.$$

However, we have already proved this in (4.28).

(b) Again $\gcd(n, k) = 1$ and $n' = n$. The case when $g = 1$ is trivial as then we have $g_x = 1$ for all $x \in \mathbb{Z}_n$ and $v_{k,n} = 0$.

Now assume $g > 1$. Since $k^g = 1 \pmod{n}$, g_1 divides g. If $g_1 < g$, then $g_1 \leq g/p_1$ which implies that $k^{g_1} \leq k^{g/p_1} = (sn + 1)^{1/p_1} = o(n)$, which is a contradiction. Thus $g = g_1$.

Now Lemma 1.4.4(c) immediately yields

$$\frac{v_{k,n}}{n} < \frac{2k^{g_1/p_1}}{n} \leq \frac{2(1 + sn)^{1/p_1}}{n} = o(1). \qquad \square$$

Now we state LSD results for two types of k-circulant matrices with dependent inputs.

Type I. $n = k^g + 1$ for some fixed $g \geq 2$. We observe a simple but crucial property of the eigenvalue partitioning $\{\mathcal{P}_j\}$ of $\mathbb{Z}_n$ (see (1.7) of Chapter 1). For every integer $t \geq 0$, $tk^g = (-1 + n)t = -t \pmod{n}$. Hence λ_t and λ_{n-t} belong to the same partition block $S(t) = S(n - t)$. Thus each $S(t)$ contains an even number of elements, except for $t = 0, \frac{n}{2}$. Hence the eigenvalue partitioning sets $\mathcal{P}_j$ are self-conjugate. So, we can find sets $\mathcal{A}_j \subset \mathcal{P}_j$ such that

$$\mathcal{P}_j = \{x : x \in \mathcal{A}_j \text{ or } n - x \in \mathcal{A}_j\} \text{ and } \#\mathcal{A}_j = \frac{1}{2}\#\mathcal{P}_j. \qquad (4.29)$$

However, it follows from Lemma 4.5.2 that for $n = k^g + 1$, $g_1 = 2g$ and $\upsilon_{k,n}/n \to 0$. Indeed, for $g = 2$, it is easy to check that $S(1) = \{1, k, n-1, n-k\}$, and hence $g_1 = 4$. Thus

$$\{x \in \mathbb{Z}_n : g_x < g_1\} = \begin{cases} \{0, n/2\} & \text{if } n \text{ is even} \\ \{0\} & \text{if } n \text{ is odd.} \end{cases} \quad (4.30)$$

As a consequence, $\upsilon_{k,n}/n \le 2/n \to 0$.

For any $d \ge 1$ and $\{E_i\}$ i.i.d. Exp(1), let

$$G_d(x) = \mathrm{P}\left(\prod_{i=1}^{d} E_i \le x\right).$$

Note that G_d is continuous. For any integer $d \ge 1$, define $H_d(s_1, \ldots, s_d, x)$ for $(s_1, s_2, \ldots, s_d) \in [0, 2\pi]^d$ and $x \ge 0$, as

$$H_d(s_1, \ldots, s_d, x) = \begin{cases} G_d\left(\dfrac{x^{2d}}{(2\pi)^d \prod_{i=1}^{d} f(s_i)}\right) & \text{if } \prod_{i=1}^{d} f(s_i) \ne 0 \\ 1 & \text{if } \prod_{i=1}^{d} f(s_i) = 0. \end{cases}$$

The proof of the following lemma is omitted.

Lemma 4.5.3. (a) For fixed x, $H_d(s_1, \ldots, s_d, x)$ is bounded continuous on $[0, 2\pi]^d$.

(b) F_d defined below is a valid continuous distribution function.

$$F_d(x) = \int_0^1 \cdots \int_0^1 H_d(2\pi t_1, \ldots, 2\pi t_d, x) \prod_{i=1}^{d} dt_i \quad \text{for } x \ge 0. \quad (4.31)$$

Theorem 4.5.4 (Bose et al. (2009)). Suppose Assumptions 4.1.1 and 4.1.2 hold. Suppose $n = k^g + 1$ for some fixed $g \ge 2$. Then as $n \to \infty$, $F_{n^{-1/2}A_{k,n}}$ converges in L_2 to the distribution of $U_1(\prod_{i=1}^{g} E_i)^{1/2g}$ where $\{E_i\}$ are i.i.d. with distribution function F_g given in (4.31) and U_1 is uniformly distributed over the $(2g)$-th roots of unity, independent of the $\{E_i\}$.

Remark 4.5.1. If $\{x_i\}$ are i.i.d., then $f(s) = 1/2\pi$ for all $s \in [0, 2\pi]$ and the LSD is $U_1(\prod_{i=1}^{g} E_i)^{1/2g}$ where $\{E_i\}$ are i.i.d. Exp(1), U_1 is as in Theorem 4.5.4 and independent of $\{E_i\}$. This limit agrees with Theorem 3.3.3(a).

Proof of Theorem 4.5.4. This proof also uses the normal approximation, and the eigenvalue description given in Section 1.4 of Chapter 1.

For simplicity we first prove the result when $g = 2$. Note that $gcd(k, n) = 1$, and hence in this case $n' = n$ in (1.9) of Chapter 1. Recall that $\upsilon_{k,n}$ is the total number of eigenvalues γ_j of $A_{k,n}$ such that $j \in \mathcal{P}_l$ and $|\mathcal{P}_l| < g_1$. In view of Lemma 4.5.2(a), we have $\upsilon_{k,n}/n \to 0$, and hence these eigenvalues do not contribute to the LSD. Hence it remains to consider *only* the eigenvalues corresponding to the sets $\mathcal{P}_l$ which have size *exactly equal* to g_1.

Note that $S(1) = \{1, k, n-1, n-k\}$, and hence $g_1 = 4$. Recall the quantities $n_j = \#\mathcal{P}_j$, $y_j = \prod_{t \in \mathcal{P}_j} \lambda_t$, where λ_j, $0 \leq j < n$ are as in (1.2) of Chapter 1. Also, for every integer $t \geq 0$, $tk^2 = -t \pmod{n}$, so that λ_t and λ_{n-t} belong to the same partition block $S(t) = S(n-t)$. Thus each y_t is real. Let us define

$$I_n = \{l : \#\mathcal{P}_l = 4\}.$$

It is clear that $\frac{n}{\#I_n} \to 4$. Without any loss, let $I_n = \{1, 2, \ldots, \#I_n\}$.

Let $1, \omega, \omega^2, \omega^3$ be the fourth roots of unity. Now, for every j, the eigenvalues of $n^{-1/2}A_{k,n}$ corresponding to the set $\mathcal{P}_j$ are: $\left(\frac{y_j}{n^2}\right)^{1/4}\omega^k$, $k = 0, 1, \ldots, 3$. Hence it suffices to consider only the modulus $\left(\frac{y_j}{n^2}\right)^{1/4}$ as j varies: if these have an LSD F, say, then the LSD of the whole sequence will be (r, θ) in polar coordinates where r is distributed according to F and θ is distributed uniformly across all the fourth roots of unity. Moreover r and θ are independent. With this in mind, we consider for $x > 0$,

$$F_n(x) = \frac{1}{\#I_n} \sum_{i=1}^{\#I_n} \mathbb{I}\left(\left[\frac{y_j}{n^2}\right]^{\frac{1}{4}} \leq x\right).$$

Since the set of λ values corresponding to any $\mathcal{P}_j$ is closed under conjugation, there exists a set $\mathcal{A}_i \subset \mathcal{P}_i$ of size 2 (see (4.29)) such that

$$\mathcal{P}_i = \{x : x \in \mathcal{A}_i \text{ or } n - x \in \mathcal{A}_i\}.$$

Combining each λ_j with its conjugate, we may write y_j in the form,

$$y_j = \prod_{t \in \mathcal{A}_j} (b_t^2 + c_t^2), \tag{4.32}$$

where

$$b_t = \sum_{j=0}^{n-1} x_j \cos(\omega_t j) \quad \text{and} \quad c_t = \sum_{j=0}^{n-1} x_j \sin(\omega_t j).$$

Note that for $x > 0$,

$$E[F_n(x)] = \frac{1}{\#I_n} \sum_{j=1}^{\#I_n} P\left(\frac{y_j}{n^2} \leq x^4\right).$$

So our first aim is to show

$$\frac{1}{\#I_n} \sum_{j=1}^{\#I_n} P\left(\frac{y_j}{n^2} \leq x^4\right) \to F_2(x),$$

where $F_2(x)$ is as in (4.31) with $d = 2$. Now using (4.32) and Theorem 4.2.4, we can write $\frac{y_j}{n^2}$ as

$$\frac{y_j}{n^2} = L_{n,j} + R_{n,j} \text{ for } 1 \leq j \leq \#I_n,$$

where

$$L_{n,j} = 4\pi^2 f_j \bar{y}_j, \quad \bar{y}_j = \prod_{t \in A_j} (\xi_{2t-1}^2 + \xi_{2t}^2), \quad f_j = \prod_{t \in A_j} f(\omega_t), \quad (4.33)$$

and $\{\xi_t\}$ are as in Theorem 4.2.4. If $A_j = \{j_1, j_2\}$ then

$$R_{n,j} = L_n(\omega_{j_1})R_n(\omega_{j_2}) + L_n(\omega_{j_2})R_n(\omega_{j_1}) + R_n(\omega_{j_1})R_n(\omega_{j_2}),$$

where $L_n(\omega_t)$ and $R_n(\omega_t)$ are as in (4.11). Now using Theorem 4.2.4 it is easy to see that for any $\epsilon > 0$, as $n \to \infty$, we have $\max_{1 \leq j \leq \#I_n} E(|R_{n,j}| > \epsilon) \to 0$. So in view of Lemma 4.2.3 it is enough to show that

$$\frac{1}{\#I_n} \sum_{j=1}^{\#I_n} P\left(L_{n,j} \leq x^4\right) \to F_2(x). \quad (4.34)$$

We show this in two steps.

Step I. Normal approximation:

$$\frac{1}{\#I_n} \sum_{j=1}^{\#I_n} \left[P\left(L_{n,j} \leq x^4\right) - \Phi_4(A_{n,j}) \right] \to 0 \quad \text{as } n \to \infty, \quad (4.35)$$

where

$$A_{n,j} = \left\{ (x_1, y_1, x_2, y_2) \in \mathbb{R}^4 : \prod_{i=1}^{2} [2^{-1}(x_i^2 + y_i^2)] \leq \frac{x^4}{4\pi^2 f_j} \right\}, \quad 1 \leq j \leq \#I_n.$$

Step II. $\displaystyle \lim_{n \to \infty} \frac{1}{\#I_n} \sum_{j=1}^{\#I_n} \Phi_4(A_{n,j}) = F_2(x).$

Proof of Step I. It is important to note that $A_{n,j}$ is *non-convex*. So we have to apply care while using normal approximation. Define

$$X_{l,j} = 2^{1/2} \left(\epsilon_l \cos\left(\frac{2\pi t l}{n}\right), \epsilon_l \sin\left(\frac{2\pi t l}{n}\right), \ t \in A_j \right), \ 0 \leq l < n, 1 \leq j \leq \#I_n.$$

Note that $\{X_{l,j}\}$ satisfies (4.14)–(4.16), and

$$\left\{ L_{n,j} \leq x^4 \right\} = \left\{ \frac{1}{\sqrt{n}} \sum_{l=1}^{n-1} X_{l,j} \in A_{n,j} \right\}.$$

For (4.35), it suffices to show that for every $\epsilon > 0$, there exists $N = N(\epsilon)$ such that for all $n \geq N(\epsilon)$,

$$\sup_{j \in I_n} \left| P\left(L_{n,j} \leq x^4\right) - \Phi_4(A_{n,j}) \right| \leq \epsilon.$$

Fix $\epsilon > 0$ and $M_1 > 0$ so that $\Phi([-M_1, M_1]^c) \le \epsilon/16$. By Assumption 4.1.2,

$$\mathrm{E}\left(\frac{1}{\sqrt{n}}\sum_{l=0}^{n-1}\varepsilon_l \cos\frac{2\pi l t}{n}\right)^2 = \mathrm{E}\left(\frac{1}{\sqrt{n}}\sum_{l=0}^{n-1}\varepsilon_l \sin\frac{2\pi l t}{n}\right)^2 = 1/2,$$

for any $n \ge 1$ and $0 < t < n$. Now by Chebyshev's inequality, we can find $M_2 > 0$ such that for each $n \ge 1$ and for each $0 < t < n$,

$$\mathrm{P}\left(|\frac{1}{\sqrt{n}}\sum_{l=0}^{n-1}\varepsilon_l \cos\frac{2\pi l t}{n}| \ge M_2\right) \le \epsilon/16.$$

The same bound holds for the sine term also. Set $M = \max\{M_1, M_2\}$. Define the set

$$B := \left\{(x_1, y_1, x_2, y_2) \in \mathbb{R}^4 : |x_j|, |y_j| \le M \ \ \forall\, j\right\}.$$

Then for all $j \in I_n$,

$$\left|\mathrm{P}\left(\frac{1}{\sqrt{n}}\sum_{l=0}^{n-1}X_{l,j} \in A_{n,j}\right) - \Phi_4(A_{n,j})\right|$$

$$\le \left|\mathrm{P}\left(\frac{1}{\sqrt{n}}\sum_{l=0}^{n-1}X_{l,j} \in A_{n,j} \cap B\right) - \Phi_4(A_{n,j} \cap B)\right| + \epsilon/2.$$

Since $A_{n,j}$ is a *non-convex* set, we now apply Lemma 3.1.1(b) for $A_{n,j} \cap B$ to obtain

$$\sup_{j \in I_n}\left|\mathrm{P}\left(\frac{1}{\sqrt{n}}\sum_{l=0}^{n-1}X_{l,j} \in A_{n,j} \cap B\right) - \Phi_4(A_{n,j} \cap B)\right|$$

$$\le C_1 n^{-\delta/2}\rho_{2+\delta} + 2\sup_{j \in I_n}\sup_{z \in \mathbb{R}^4}\Phi_4\left((\partial(A_{n,j} \cap B))^\eta - z\right),$$

where

$$\rho_{2+\delta} = \sup_{j \in I_n}\frac{1}{n}\sum_{l=0}^{n-1}\mathrm{E}\,\|X_{l,j}\|^{2+\delta} \quad \text{and} \quad \eta = \eta(n) = C_2\rho_{2+\delta}n^{-\delta/2}.$$

Note that $\rho_{2+\delta}$ is uniformly bounded in n by Assumption 4.1.2.

It thus remains to show that for all sufficiently large n,

$$\sup_{j \in I_n}\sup_{z \in \mathbb{R}^4}\Phi_4\left((\partial(A_{n,j} \cap B))^\eta - z\right) \le \epsilon/8. \tag{4.36}$$

Note that $\partial(A_{n,j} \cap B) \subseteq \partial A_{n,j} \cap \partial B \subseteq \partial B$, and hence

$$
\sup_{j \in I_n} \sup_{z \in \mathbb{R}^4} \Phi_{2g}\left((\partial(A_{n,j} \cap B))^\eta - z \right)
$$

$$
= \sup_{j \in I_n} \sup_{z \in \mathbb{R}^4} \int_{(\partial(A_{n,j} \cap B))^\eta} \phi(x_1 - z_1) \cdots \phi(y_2 - z_4) dx_1 \ldots dy_2
$$

$$
\leq \sup_{z \in \mathbb{R}^4} \int_{(\partial B)^\eta} \phi(x_1 - z_1) \cdots \phi(y_2 - z_4) dx_1 \ldots dy_2
$$

$$
\leq \int_{(\partial B)^\eta} dx_1 \ldots dy_2.
$$

Finally note that ∂B is a *compact* 3-dimensional manifold which has zero measure under the 4-dimensional Lebesgue measure. By compactness of ∂B, we have $(\partial B)^\eta \downarrow \partial B$ as $\eta \to 0$, and (4.36) follows by dominated convergence theorem. Therefore $\mathrm{E}[F_n(x)]$ equals

$$
\frac{1}{\#I_n} \sum_{j=1}^{\#I_n} \mathrm{P}\left(\frac{y_j}{n^2} \leq x^4 \right) \sim \frac{1}{\#I_n} \sum_{j=1}^{\#I_n} \mathrm{P}\left(L_{n,j} \leq x^4 \right) \sim \frac{1}{\#I_n} \sum_{j=1}^{\#I_n} \Phi_4(A_{n,j}).
$$

Proof of Step II. To identify the limit, recall the structure of the sets $S(x), \mathcal{P}_j, \mathcal{A}_j$ and their properties. Since $\#I_n/n \to 1/4$, $v_{k,n} \leq 2$ and either $S(x) = S(u)$ or $S(x) \cap S(u) = \emptyset$, we have

$$
\lim_{n \to \infty} \frac{1}{\#I_n} \sum_{j=1}^{\#I_n} \Phi_4(A_{n,j}) = \lim_{n \to \infty} \frac{1}{n} \sum_{j=1, |\mathcal{A}_j|=2}^{n} \Phi_4(A_{n,j}). \tag{4.37}
$$

For $n = k^2 + 1$, write $\{1, 2, \ldots, n-1\}$ as $\{ak + b;\ 0 \leq a \leq k-1,\ 1 \leq b \leq k\}$. Using the construction of $S(x)$, (except for at most two values of j),

$$
A_j = \{ak + b, bk - a\} \text{ for } j = ak + b;\ 0 \leq a \leq k-1,\ 1 \leq b \leq k.
$$

Recall that for fixed x, $H_2(s_1, s_2, x)$ is uniformly continuous on $[0, 2\pi] \times [0, 2\pi]$. Therefore, given any positive number ρ, we can choose N large enough such that for all $n = k^2 + 1 > N$,

$$
\sup_{0 \leq a \leq k-1,\ 1 \leq b \leq k} \left| H_2\left(\frac{2\pi(ak + b)}{n}, \frac{2\pi(bk - a)}{n}, x \right) - H_2\left(\frac{2\pi a}{\sqrt{n}}, \frac{2\pi b}{\sqrt{n}}, x \right) \right| < \rho. \tag{4.38}
$$

Finally, using (4.37) and (4.38) we have

$$
\begin{aligned}
\lim_{n\to\infty} \frac{1}{\#I_n} \sum_{j=1}^{\#I_n} \Phi_4(A_{n,j}) &= \lim_{n\to\infty} \frac{1}{n} \sum_{j=1}^{n} \Phi_4(A_{n,j}) \\
&= \lim_{n\to\infty} \frac{1}{n} \sum_{j=1}^{n} G_2\left(\frac{x^4}{4\pi^2 f_j}\right) \\
&= \lim_{n\to\infty} \frac{1}{n} \sum_{b=1}^{\lfloor\sqrt{n}\rfloor} \sum_{a=0}^{\lfloor\sqrt{n}\rfloor} H_2\left(\frac{2\pi(ak+b)}{n}, \frac{2\pi(bk-a)}{n}, x\right) \\
&= \lim_{n\to\infty} \frac{1}{n} \sum_{b=1}^{\lfloor\sqrt{n}\rfloor} \sum_{a=0}^{\lfloor\sqrt{n}\rfloor} H_2\left(\frac{2\pi a}{\sqrt{n}}, \frac{2\pi b}{\sqrt{n}}, x\right) \\
&= \int_0^1 \int_0^1 H_2(2\pi s, 2\pi t, x)\, ds\, dt \\
&= F_2(x).
\end{aligned}
$$

To show that $\mathrm{Var}[F_n(x)] \to 0$, since the variables involved are all bounded, it is enough to show that

$$
n^{-2} \sum_{j\neq j'} \mathrm{Cov}\left(\mathbb{I}\left(\frac{y_j}{n^2} \leq x^4\right), \mathbb{I}\left(\frac{y_{j'}}{n^2} \leq x^4\right)\right) \to 0.
$$

This can be shown by extending the proof used to show $E[F_n(x)] \to F_2(x)$. We have to extend the vectors with 4 coordinates defined above to ones with 8 coordinates and proceed. We omit the details. This completes the proof of the Theorem for $g = 2$.

The above argument can be extended to cover the general ($g > 2$) case. We highlight only a few of the technicalities and omit the other details. For general g we need the following lemma.

Lemma 4.5.5. Suppose $L_n(\omega_j)$, $R_n(\omega_j)$ are as defined in Theorem 4.2.4. Then given any $\epsilon, \eta > 0$, there exists an $N \in \mathbb{N}$ such that

$$
P\left(\left|\prod_{i=1}^{s} L_n(\omega_{j_i}) \prod_{i=s+1}^{g} R_n(\omega_{j_i})\right| > \epsilon\right) < \eta \text{ for all } n \geq N.
$$

Proof. Note that by an iterative argument,

$$\mathrm{P}\left(\left|\prod_{i=1}^{s} L_n(\omega_{j_i}) \prod_{i=s+1}^{g} R_n(\omega_{j_i})\right| > \epsilon\right) \leq \mathrm{P}\left(\left|\prod_{i=2}^{s} L_n(\omega_{j_i}) \prod_{i=s+1}^{g} R_n(\omega_{j_i})\right| > 1/M\right)$$

$$+ \mathrm{P}\left(\left|L_n(\omega_{j_1})\right| \geq M\epsilon\right)$$

$$\leq \sum_{i=2}^{s} \mathrm{P}\left(\left|L_n(\omega_{j_i})\right| \geq M\right)|$$

$$+ \mathrm{P}\left(\left|\prod_{i=s+1}^{g} R_n(\omega_{j_i})\right| > 1/M^s\right)$$

$$+ \mathrm{P}\left(\left|L_n(\omega_{j_1})\right| \geq M\epsilon\right).$$

Further note that

$$\mathrm{P}\left(\left|\prod_{i=s+1}^{g} R_n(\omega_{j_i})\right| > 1/M^s\right)$$

$$\leq \mathrm{P}\left(\left|\prod_{i=s+2}^{g} R_n(\omega_{j_i})\right| > 1/M^s\right) + \mathrm{P}\left(\left|R_n(\omega_{j_{s+1}})\right| > 1\right)$$

$$\leq \mathrm{P}\left(\left|R_n(\omega_{j_g})\right| > 1/M^s\right) + \sum_{i=s+1}^{g-1} \mathrm{P}\left(\left|R_n(\omega_{j_i})\right| > 1\right)$$

$$\leq \left(M^s + g - s - 1\right) \max_{1 \leq k \leq n} \mathrm{E}\left|R_n(\omega_k)\right|.$$

Combining all the above, we get

$$\mathbb{P}\left(\left|\prod_{i=1}^{s} L_n(\omega_{j_i}) \prod_{i=s+1}^{g} R_n(\omega_{j_i})\right| > \epsilon\right)$$

$$\leq \mathrm{P}\left(\left|L_n(\omega_{j_1})\right| \geq M\epsilon\right) + \sum_{i=2}^{s} \mathrm{P}\left(\left|L_n(\omega_{j_i})\right| \geq M\right)$$

$$+ \left(M^s + g - s - 1\right) \max_{1 \leq k \leq n} \mathrm{E}\left|R_n(\omega_k)\right|$$

$$\leq \frac{1}{M}(s - 1 + 1/\epsilon)4\pi \max_{s \in [0, 2\pi]} f(s) + \left(M^s + g - s - 1\right) \max_{1 \leq k \leq n} \mathrm{E}\left|R_n(\omega_k)\right|.$$

The first term in the right side is smaller than $\eta/2$ if we choose M large. Since $\max_{1 \leq k \leq n} \mathrm{E}\left|R_n(\omega_k)\right| \to 0$ as $n \to \infty$, we can choose $N \in \mathbb{N}$ such that the second term is less than $\eta/2$ for all $n \geq N$, proving the lemma. $\square$

Now we return to the main proof for general $g \geq 2$. As before, $n' = n$ and $\upsilon_{k,n}/n \to 0$. Hence it remains to consider only the eigenvalues corresponding to the sets $\mathcal{P}_l$ which have size exactly equal to g_1. It follows from Lemma

4.5.2(a) that $g_1 = 2g$. We can now proceed as in the case of $g = 2$. First we show that

$$\frac{1}{\#I_n} \sum_{j=1}^{\#I_n} \mathrm{P}\left(\frac{y_j}{n^g} \leq x^{2g}\right) \to F_g(x). \tag{4.39}$$

Now write $\frac{y_j}{n^g}$ as

$$\frac{y_j}{n^g} = L_{n,j} + R_{n,j} \text{ for } 1 \leq j \leq \#I_n,$$

where

$$L_{n,j} = \prod_{t \in \mathcal{A}_j} L_n(\omega_t) = (2\pi)^g f_j \frac{y_j}{n^g},$$

f_j and $\bar{y}_j$ are as in (4.33). Using Lemma 4.5.5, it is easy to show that $\max_{1 \leq j \leq \#I_n} \mathrm{P}(|R_{n,j}| > \epsilon) \to 0$, for any $\epsilon > 0$. So, by Lemma 4.2.3, to show (4.39), it is sufficient to show that

$$\frac{1}{\#I_n} \sum_{j=1}^{\#I_n} \mathrm{P}\left(L_{n,j} \leq x^{2g}\right) \to F_g(x).$$

We prove this in two steps as we did for $g = 2$. Define

$$\bar{A}_{n,j} = \left\{(x_i, y_i, i = 1, 2, \ldots, g) \in \mathbb{R}^{2g} : \prod_{i=1}^{g}[2^{-1}(x_i^2 + y_i^2)] \leq \frac{x^{2g}}{(2\pi)^g f_j}\right\}.$$

Now, in Step I, for fixed $\epsilon > 0$ we find $M_1 > 0$ large such that $\Phi([-M_1, M_1]^c) \leq \epsilon/(8g)$ and $M_2 > 0$ such that

$$\mathrm{P}\left(|\frac{1}{\sqrt{n}} \sum_{l=0}^{n-1} \varepsilon_l \cos \frac{2\pi lt}{n}| \geq M_2\right) \leq \epsilon/(8g),$$

and the same bound holds for the sine term. Set $M = \max\{M_1, M_2\}$ and define

$$B := \left\{(x_j, y_j; 1 \leq j \leq g) \in \mathbb{R}^{2g} : |x_j|, |y_j| \leq M \ \forall \ j\right\}.$$

Note that, ∂B is a *compact* $(2g-1)$-dimensional manifold which has zero measure under the $2g$-dimensional Lebesgue measure. Now proceeding as before, we have

$$\left|\frac{1}{\#I_n} \sum_{l=1}^{\#I_n} \mathrm{P}\left(L_{n,j} \leq x^4\right) - \frac{1}{\#I_n} \sum_{l=1}^{\#I_n} \Phi_4(\bar{A}_{n,j})\right| \to 0.$$

Now note that for $n = k^g + 1$ we can write $\{1, 2, \ldots, n-1\}$ as $\{b_1 k^{g-1} + b_2 k^{g-2} + \cdots + b_{g-1}k + b_g; 0 \leq b_i \leq k - 1, \text{ for } 1 \leq i \leq k - 1; 1 \leq b_g \leq k\}$. So we can write the sets $\mathcal{A}_j$ (see (4.29)) explicitly using this decomposition of

$\{1, 2, \ldots, n-1\}$ as done in the case $g = 2$, that is, in the case $n = k^2 + 1$. For example, if $g = 3$,

$$\mathcal{A}_j = \{b_1 k^2 + b_2 k + b_3, \ b_2 k^2 + b_3 k - b_1, \ b_3 k^2 - b_1 k - b_2\},$$

for $j = b_1 k^2 + b_2 k + b_3$ (except for finitely many j, bounded by $v_{k,n}$ and they do not contribute to the limit). Using this fact, and proceeding as before, we conclude that the LSD is now $F_g(\cdot)$, proving Theorem 4.5.4 completely. □

Type II. $n = k^g - 1$ for some $g \geq 2$. First we extend the definition of $B(s)$ given in (4.4) and define $B(s_1, \ldots, s_g)$ as

$$\begin{pmatrix} \psi_1(e^{is_1}) & -\psi_2(e^{is_1}) & 0 & 0 & \cdots & 0 \\ \psi_2(e^{is_1}) & \psi_1(e^{is_1}) & 0 & 0 & \cdots & 0 \\ 0 & 0 & \psi_1(e^{is_2}) & -\psi_2(e^{is_2}) & \cdots & 0 \\ 0 & 0 & \psi_2(e^{is_2}) & \psi_1(e^{is_2}) & \cdots & 0 \\ 0 & 0 & 0 & 0 & \vdots & 0 \\ 0 & 0 & 0 & \cdots & \psi_1(e^{is_g}) & -\psi_2(e^{is_g}) \\ 0 & 0 & 0 & \cdots & \psi_2(e^{is_g}) & \psi_1(e^{is_g}) \end{pmatrix}.$$

For $z_i, w_i \in \mathbb{R}, i = 1, 2, \ldots, g$, and with $\{N_i\}$ i.i.d. $N(0, 1)$, define

$$\mathcal{H}_g(s_i, z_i, w_i, i = 1, \ldots, g)$$
$$= P\left(B(s_1, \ldots, s_g)(N_1, \ldots, N_{2g})' \leq (z_i, w_i, i = 1, \ldots, g)'\right).$$

Let $C_0 = \{t \in [0, 1] : f(2\pi t) = 0\}$. The proof of the following lemma is omitted.

Lemma 4.5.6. (a) For fixed $\{z_i, w_i, i = 1, \ldots, g\}$, $\mathcal{H}_g$ is a bounded continuous function of $(s_1, \ldots, s_g) \in [0, 2\pi]^g$.

(b) $\mathcal{F}_g$ defined below is a proper distribution function.

$$\mathcal{F}_g(z_i, w_i, i = 1, \ldots, g) = \int_0^1 \cdots \int_0^1 \mathcal{H}_g(2\pi t_i, z_i, w_i, i = 1, \ldots, g) \prod dt_i.$$
(4.40)

(c) If $\mathrm{Leb}(C_0) = 0$ then $\mathcal{F}_g(z_i, w_i, i = 1, .., g)$ is continuous everywhere and equals

$$\int \mathbb{I}_{\{t \leq (z_k, w_k, k=1,..,g)\}} \int_0^1 \cdots \int_0^1 \frac{\mathbb{I}_{\{\prod_{i=1}^g f(2\pi u_i) \neq 0\}}}{(2\pi)^g \prod_{i=1}^g [\pi f(2\pi u_i)]} \prod_{i=1}^g e^{-\frac{1}{2}\frac{t_{2i-1}^2 + t_{2i}^2}{\pi f(2\pi u_i)}} d\mathbf{u} \, d\mathbf{t},$$

where $\mathbf{t} = (t_1, \ldots, t_{2g})$, $d\mathbf{t} = \prod dt_i$, and $d\mathbf{u} = \prod du_i$. $\mathcal{F}_g$ is multivariate normal if and only if f is constant almost everywhere (Lebesgue).

(d) If $\mathrm{Leb}(C_0) \neq 0$ then $\mathcal{F}_g$ is discontinuous only on $D_g = \{(z_i, w_i, i = 1, \ldots, g) : \prod_{i=1}^g z_i w_i = 0\}$.

Theorem 4.5.7 (Bose et al. (2009)). Suppose Assumptions 4.1.1 and 4.1.2 hold. Suppose $n = k^g - 1$ for some $g \geq 2$. Then as $n \to \infty$, $F_{n^{-1/2} A_{k,n}}$ converges in L_2 to the distribution of $(\prod_{i=1}^{g} G_i)^{1/g}$ where $(\mathcal{R}(G_i), \mathcal{I}(G_i); \ i = 1, 2, \ldots g)$ has the distribution $\mathcal{F}_g$ given in (4.40).

Remark 4.5.2. If $\{x_i\}$ are i.i.d., with finite $(2 + \delta)$-th moment, then $f(s) \equiv 1/2\pi$ and the limit simplifies to $U_2(\prod_{i=1}^{g} E_i)^{1/2g}$ where $\{E_i\}$ are i.i.d. $Exp(1)$ and U_2 is uniformly distributed over the unit circle independent of $\{E_i\}$. This agrees with the conclusion in Theorem 3.3.3(b).

Proof of Theorem 4.5.7. First assume $\text{Leb}(C_0) = 0$. Since $k^g = 1 + n = 1 \bmod n$, we have $\gcd(k, n) = 1$ and $g_1 | g$. If $g_1 < g$, then $g_1 \leq g/\alpha$ where $\alpha = 2$ if g is even and $\alpha = 3$ if g is odd. In either case,

$$
\begin{aligned}
k^{g_1} &\leq k^{g/\alpha} \\
&\leq (1 + n)^{1/\alpha} = o(n).
\end{aligned}
$$

Hence $g = g_1$. By Lemma 4.5.2(b), the total number of eigenvalues γ_j of $A_{k,n}$ such that, $j \in \mathcal{A}_l$ and $|\mathcal{A}_l| < g$, is asymptotically negligible.

Unlike in the previous theorem, here the partition sets $\mathcal{A}_l$ are not necessarily self-conjugate. However, the number of indices l such that $\mathcal{A}_l$ is self-conjugate is asymptotically negligible compared to n. To show this, we need to bound the cardinality of the following set for $1 \leq l < g$:

$$
\begin{aligned}
D_l &= \{t \in \{1, 2, \ldots, n\} : tk^l = -t \ (\bmod \ n)\} \\
&= \{t \in \{1, 2, \ldots, n\} : n | t(k^l + 1)\}.
\end{aligned}
$$

Note that $t_0 = n / \gcd(n, k^l + 1)$ is the minimum element of D_l and every other element is a multiple of t_0. Thus

$$
|D_l| \leq \frac{n}{t_0} \leq \gcd(n, k^l + 1).
$$

Let us now estimate $\gcd(n, k^l + 1)$. For $l > \lceil g/2 \rceil$,

$$
\begin{aligned}
\gcd(n, k^l + 1) &\leq \gcd(k^g - 1, k^l + 1) \\
&= \gcd\left(k^{g-l}(k^l + 1) - (k^{g-l} - 1), k^l + 1\right) \\
&\leq k^{g-l},
\end{aligned}
$$

which implies $\gcd(n, k^l + 1) \leq k^{\lceil g/2 \rceil}$ for all $1 \leq l < g$. Therefore

$$
\frac{\gcd(n, k^l + 1)}{n} = \frac{k^{\lceil g/2 \rceil}}{(k^g - 1)} \leq \frac{2}{k^{\lceil (g+1)/2 \rceil}} \leq \frac{2}{((n)^{1/g})^{\lceil (g+1)/2 \rceil}} = o(1).
$$

So, we can ignore the sets $\mathcal{P}_j$ which are self-conjugate. For other $\mathcal{P}_j$, y_j defined below will be complex.

$$
y_j = \prod_{t \in \mathcal{P}_j} (b_t + ic_t).
$$

For simplicity we provide the detailed argument only for $g = 2$. Then, $n = k^2 - 1$ and write $\{0, 1, 2, \ldots, n\}$ as $\{ak + b; \ 0 \leq a \leq k-1, \ 0 \leq b \leq k-1\}$. Using the construction of $S(x)$ we have $\mathcal{P}_j = \{ak+b, bk+a\}$ and $\#\mathcal{P}_j = 2$ for $j = ak + b; \ 0 \leq a \leq k-1, \ 0 \leq b \leq k-1$ (except for finitely many j and hence such indices do not affect the LSD). Let us define

$$I_n = \{j : \#\mathcal{P}_j = 2\}.$$

Clearly, $n/\#I_n \to 2$. Without loss, let $I_n = \{1, 2, \ldots, \#I_n\}$. Suppose $\mathcal{P}_j = \{j_1, j_2\}$. We first show the L_2 convergence of the empirical distribution of $\frac{1}{\sqrt{n}}(\sqrt{n}b_{j_1}, \sqrt{n}c_{j_1}, \sqrt{n}b_{j_2}, \sqrt{n}c_{j_2})$ for those j for which $\#\mathcal{P}_j = 2$. Let $F_n(x, y, z, w)$ be the ESD of $\{(b_{j_1}, c_{j_1}, b_{j_2}, c_{j_2})\}$, that is,

$$F_n(z_1, w_1, z_2, w_2) = \frac{1}{\#I_n} \sum_{j=1}^{\#I_n} \mathbb{I}\big(b_{j_k} \leq z_k, c_{j_k} \leq w_k, \ k = 1, 2\big).$$

We show that for $z_1, w_1, z_2, w_2 \in \mathbb{R}$,

$$\mathrm{E}[F_n(z_1, w_1, z_2, w_2)] \to \mathcal{F}_2(z_1, w_1, z_2, w_2) \text{ and } \mathrm{Var}[F_n(z_1, w_1, z_2, w_2)] \to 0. \tag{4.41}$$

Let $Y_n(\omega_j)$ be as in Lemma 4.2.5 and $Y_{1n}(\omega_j) = \mathcal{R}(Y_n(\omega_j))$, $Y_{2n}(\omega_j) = \mathcal{I}(Y_n(\omega_j))$. For $j = 1, 2, \ldots, n$, let

$$\eta_j = (\xi_{2j_1-1}, \xi_{2j_1}, \xi_{2j_2-1}, \xi_{2j_2})',$$
$$Y_{n,j} = \big(Y_{1n}(\omega_{j_1}), Y_{2n}(\omega_{j_1}), Y_{1n}(\omega_{j_2}), Y_{2n}(\omega_{j_2})\big)'.$$

Then

$$(b_{j_1}, c_{j_1}, b_{j_2}, c_{j_2})' = B(\omega_{j_1}, \omega_{j_2})\eta_j + Y_{n,j}.$$

By Lemma 4.2.5, for any $\epsilon > 0$, $\max_{1 \leq j \leq n} \mathrm{P}(\|Y_{n,j}\| > \epsilon) \to 0$ as $n \to \infty$. So in view of Lemma 4.2.3, to show $\mathrm{E}[F_n(z_1, w_1, z_2, w_2)] \to \mathcal{F}_2(z_1, w_1, z_2, w_2)$, it is enough to show that

$$\frac{1}{\#I_n} \sum_{j=1}^{\#I_n} \mathrm{P}(B(\omega_{j_1}, \omega_{j_2})\eta_j \leq (z_1, w_1, z_2, w_2)') \to \mathcal{F}_2(z_1, w_1, z_2, w_2).$$

For this we use normal approximation. Let $N = (N_1, N_2, N_3, N_4)'$, where $\{N_i\}$ are i.i.d. $N(0, 1)$ and define

$$X_{l,j} = 2^{1/2}\left(\varepsilon_l \cos\left(\frac{2\pi j_1 l}{n}\right), \varepsilon_l \sin\left(\frac{2\pi j_1 l}{n}\right), \varepsilon_l \cos\left(\frac{2\pi j_2 l}{n}\right), \varepsilon_l \sin\left(\frac{2\pi j_2 l}{n}\right)\right)'.$$

Note that

$$\big\{B(\omega_{j_1}, \omega_{j_2})\eta_j \leq (z_1, w_1, z_2, w_2)'\big\}$$
$$= \big\{B(\omega_{j_1}, \omega_{j_2})(\frac{1}{\sqrt{n}} \sum_{l=0}^{n-1} X_{l,j}) \leq (\sqrt{2}z_1, \sqrt{2}w_1, \sqrt{2}z_2, \sqrt{2}w_2)'\big\}.$$

Since

$$\{(r_1, r_2, r_3, r_4) : B(\omega_{j_1}, \omega_{j_2})(r_1, r_2, r_3, r_4)' \leq (\sqrt{2}z_1, \sqrt{2}w_1, \sqrt{2}z_2, \sqrt{2}w_2)'\}$$

is a *convex* set in $\mathbb{R}^4$ and $\{X_{l,j};\ l = 0, 1, \ldots, (n-1)\}$ satisfies (4.14)–(4.16), we can show using Lemma 3.1.1(a) that, as $n \to \infty$,

$$\frac{1}{\#I_n} \sum_{j=1}^{\#I_n} \big| \mathrm{P}(B(\omega_{j_1}, \omega_{j_2})\eta_j \leq (z_1, w_1, z_2, w_2)')$$
$$- \mathrm{P}(B(\omega_{j_1}, \omega_{j_2})N \leq (\sqrt{2}z_1, \sqrt{2}w_1, \sqrt{2}z_2, \sqrt{2}w_2)') \big| \to 0.$$

Hence

$$\lim_{n\to\infty} \frac{1}{\#I_n} \sum_{j=1}^{\#I_n} \mathrm{P}(B(\omega_{j_1}\omega_{j_2})\eta_j \leq (z_1, w_1, z_2, w_2)')$$

$$= \lim_{n\to\infty} \frac{1}{\#I_n} \sum_{j=1}^{\#I_n} \mathrm{P}(B(\omega_{j_1}, \omega_{j_2})N \leq (\sqrt{2}z_1, \sqrt{2}w_1, \sqrt{2}z_2, \sqrt{2}w_2)')$$

$$= \lim_{n\to\infty} \frac{1}{n} \sum_{j=1}^{n} \mathrm{P}(B(\omega_{j_1}, \omega_{j_2})N \leq (\sqrt{2}z_1, \sqrt{2}w_1, \sqrt{2}z_2, \sqrt{2}w_2)')$$

$$= \lim_{n\to\infty} \frac{1}{n} \sum_{j=1}^{n} \mathcal{H}_2(\omega_{j_1}, \omega_{j_2}, z_1, w_1, z_2, w_2)$$

$$= \lim_{n\to\infty} \frac{1}{n} \sum_{a=0}^{\lfloor\sqrt{n}\rfloor} \sum_{b=0}^{\lfloor\sqrt{n}\rfloor} \mathcal{H}_2\Big(\frac{2\pi(ak+b)}{n}, \frac{2\pi(bk+a)}{n}, z_1, w_1, z_2, w_2\Big)$$

$$= \lim_{n\to\infty} \frac{1}{n} \sum_{a=0}^{\lfloor\sqrt{n}\rfloor} \sum_{b=0}^{\lfloor\sqrt{n}\rfloor} \mathcal{H}_2\Big(\frac{2\pi a}{\sqrt{n}}, \frac{2\pi b}{\sqrt{n}}, z_1, w_1, z_2, w_2\Big)$$

$$= \int_0^1 \int_0^1 \mathcal{H}_2(2\pi s, 2\pi t, z_1, w_1, z_2, w_2)ds\, dt = \mathcal{F}_2(z_1, w_1, z_2, w_2).$$

Similarly we can show $\mathrm{Var}[F_n(x)] \to 0$ as $n \to \infty$.

Hence, the empirical distribution of y_j for those j for which $\#\mathcal{P}_j = 2$ converges to the distribution of $\prod_{i=1}^{2} G_i$ in L_2 such that $(\mathcal{R}(G_i), \mathcal{I}(G_i);\ i = 1, 2)$ has distribution $\mathcal{F}_2$. Hence the LSD of $\frac{1}{\sqrt{n}}A_{k,n}$ is the distribution of $\big(\prod_{i=1}^{2} G_i\big)^{1/2}$, proving the result when $g = 2$ and $\mathrm{Leb}(C_0) = 0$.

If $\mathrm{Leb}(C_0) \neq 0$ then we have to show (4.41) on D_2^c (of Lemma 4.5.6). All the above steps remain valid for all $(z_i, w_i; i = 1, 2) \in D_2^c$. Hence we have our required LSD. This proves the Theorem when $g = 2$.

For $g > 2$, we can write $\{0, 1, 2, \ldots, n\}$ as $\{b_1 k^{g-1} + b_2 k^{g-2} + \cdots + b_{g-1}k + b_g;\ 0 \leq b_i \leq k-1,\ \text{for } 1 \leq i \leq k\}$. The sets $\mathcal{A}_j$ can be explicitly written down as was done for $n = k^2 - 1$. For example, if $g = 3$,

$$\mathcal{A}_j = \{b_1 k^2 + b_2 k + b_3,\ b_2 k^2 + b_3 k + b_1,\ b_3 k^2 + b_1 k + b_2\},$$

for $j = b_1 k^2 + b_2 k + b_3$ (except for finitely many j, bounded by $v_{k,n}$ and they do not contribute to the limit). Using this and proceeding as before, we obtain the limit as $\left(\prod_{i=1}^{g} G_i\right)^{1/g}$ where $(\mathcal{R}(G_i), \mathcal{I}(G_i); \ i = 1, 2, \ldots g)$ has distribution $\mathcal{F}_g$. □

Figure 4.4 provides simulations for $g = 3$ when the input sequence is an MA(3) process.

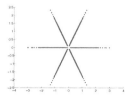 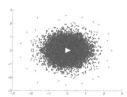

FIGURE 4.4
Eigenvalues of 10 realizations of $\frac{1}{\sqrt{n}} A_{k,n}$ where (i) (left) $n = k^3 + 1$, $k = 10$ and (ii) (right) $n = k^3 - 1$, $k = 10$. Input sequence is the MA(3) process $x_t = \varepsilon_t + 0.5\varepsilon_{t-1} + 0.3\varepsilon_{t-2} + 0.2\varepsilon_{t-3}$ where $\{\varepsilon_t\}$ is i.i.d. $N(0, 1)$.

4.6 Exercises

1. Complete the proof of Lemma 4.2.1(c).

2. Check that the variance μ_2 and the fourth moment μ_4 of F_S equal $\int_0^{1/2} 4\pi f(2\pi s)ds$ and $\int_0^{1/2} 24\pi^2 f^2(2\pi s)ds$, respectively.

3. Prove Lemma 4.4.1.

4. Suppose Assumptions 4.1.1, 4.1.2 and (4.24) hold. Show that the ESD of $\frac{1}{\sqrt{n}} PT_n$ (see (2.5) for the definition of PT_n) converges in L_2.

5. Prove Lemma 4.4.3.

6. To complete the proof of Theorem 4.3.2, show that $\text{Var}[F_n(x)] \to 0$.

7. Prove Lemma 4.5.2.

8. Find sequences $\{k = k(n)\}$ and $\{N = N(n)\}$ such that the LSD of $N^{-1/2} A_{k,N}$ has some positive mass at the origin and the rest of the probability mass is distributed as $U_1 (\prod_{i=1}^{g} E_i)^{1/2g}$ where U_1 and $\{E_i\}$ are as in Theorem 4.5.4.

5

Spectral radius: light tail

For any matrix A, its *spectral radius* $\mathrm{sp}(A)$ is defined as

$$\mathrm{sp}(A) := \max\big\{|\lambda| : \lambda \text{ is an eigenvalue of } A\big\},$$

where $|z|$ denotes the modulus of $z \in \mathbb{C}$. It has been an important object of study in random matrix theory for different types of random matrices. Our focus in this chapter is on the asymptotic behavior of this quantity for the reverse circulant, circulant and symmetric circulant random matrices when the input sequence is i.i.d.

In particular, when the input sequence has finite $(2 + \delta)$-th moment, with proper centering and scaling, the spectral radius converges to an extreme value distribution.

The spectral radius of the k-circulant is discussed in the next chapter, and of the circulant-type matrices when the input sequence is an appropriate linear process as discussed in Chapter 7.

5.1 Circulant and reverse circulant

Extreme value theory is an important area of statistics and probability. See Resnick (1987) for an excellent introduction to this area. One of the most basic distributions in this theory is the following.

Definition 5.1.1. The Gumbel distribution with parameter $\theta > 0$ is characterized by its cumulative distribution function

$$\Lambda_\theta(x) = \exp\{-\theta \exp(-x)\}, \quad x \in \mathbb{R}.$$

The special case of $\Lambda \equiv \Lambda_1$ is known as the standard Gumbel distribution.

If X_i are i.i.d. random variables then $\max_{1 \le i \le n} X_i$, after appropriate centering and scaling, converges to a Gumbel distribution, provided the tail of the distribution of X_1 satisfies certain conditions. The scaling and centering constants depend on the behavior of the tail. We shall see shortly that the Gumbel distribution arises also in connection with the asymptotic distributions of the spectral radius of the circulant-type matrices.

Recall the formulae (1.2) and (1.3) from Chapter 1 for the eigenvalues of C_n and RC_n, respectively. From, these formulae, it is clear that

$$\mathbf{sp}(C_n) = \mathbf{sp}(RC_n).$$

Hence the spectral radii for these two matrices do not have to be dealt with separately. We denote *convergence in distribution* and *convergence in probability* by $\xrightarrow{\mathcal{D}}$ and $\xrightarrow{\mathcal{P}}$, respectively.

Theorem 5.1.1 (Bose et al. (2011c)). Consider RC_n and C_n where the input $\{x_i\}$ is i.i.d. with mean μ, variance one, and $\mathrm{E}\,|x_i|^{2+\delta} < \infty$ for some $\delta > 0$.

(a) If $\mu \neq 0$ then

$$\frac{\mathbf{sp}(RC_n) - |\mu|n}{\sqrt{n}} \xrightarrow{\mathcal{D}} N(0, 1).$$

(b) If $\mu = 0$ then

$$\frac{\mathbf{sp}(\frac{1}{\sqrt{n}} RC_n) - d_q}{c_q} \xrightarrow{\mathcal{D}} \Lambda,$$

where $q = q(n) = \lfloor \frac{n-1}{2} \rfloor$, $d_q = \sqrt{\ln q}$, $c_q = \frac{1}{2\sqrt{\ln q}}$, and Λ is the standard Gumbel distribution.

Conclusions (a) and (b) hold for C_n also.

Proof. As pointed out earlier, since $\mathbf{sp}(C_n) = \mathbf{sp}(RC_n)$, it is enough to deal with only RC_n. Fortunately, the asymptotic behavior of the maximum of the periodogram has been discussed in the literature, for example by Davis and Mikosch (1999). We take help from these results.

Let $\lambda_0, \lambda_1, \ldots, \lambda_{n-1}$ be the eigenvalues of $\frac{1}{\sqrt{n}} RC_n$. Note that $\{|\lambda_k|^2; 1 \leq k < n/2\}$ is the periodogram of $\{x_i\}$ at the frequencies $\{\frac{2\pi k}{n}; 1 \leq k < n/2\}$.

If $\mu = 0$ then Theorem 2.1 of Davis and Mikosch (1999) (stated as Theorem 11.3.2 in Appendix) yields

$$\max_{1 \leq k < \frac{n}{2}} I_{x,n}(\omega_k) - \ln q \xrightarrow{\mathcal{D}} \Lambda, \tag{5.1}$$

where

$$I_{x,n}(\omega_k) = \frac{1}{n} \Big| \sum_{t=0}^{n-1} x_t e^{-it\omega_k} \Big|^2 \quad \text{and} \quad \omega_k = \frac{2\pi k}{n}.$$

Therefore

$$\max_{1 \leq k < n/2} |\lambda_k|^2 - \ln q \xrightarrow{\mathcal{D}} \Lambda,$$

and as a consequence,

$$\frac{\max_{1 \leq k < n/2} |\lambda_k|^2}{\ln q} \xrightarrow{\mathcal{P}} 1.$$

Now making the transformation $g(x) = \sqrt{x}$ and using the above two facts,

$$\frac{g(\max_{1 \leq k < n/2} |\lambda_k|^2) - g(\ln q)}{g'(\ln q)} = \frac{g'(\xi_n)}{g'(\ln q)} \left(\max_{1 \leq k < n/2} |\lambda_k|^2 - \ln q \right) \xrightarrow{\mathcal{D}} \Lambda.$$

So if $\mu = 0$, then

$$\frac{\max_{1 \leq k < \frac{n}{2}} |\lambda_k| - \sqrt{\ln q}}{\frac{1}{2\sqrt{\ln q}}} \xrightarrow{\mathcal{D}} \Lambda. \tag{5.2}$$

Now suppose that the mean of $\{x_i\}$ is $\mu > 0$. For $1 \leq k < n/2$,

$$|\lambda_k| = \frac{1}{\sqrt{n}} \Big| \sum_{t=0}^{n-1} x_t e^{it\omega_k} \Big| = \frac{1}{\sqrt{n}} \Big| \sum_{t=0}^{n-1} (x_t - \mu) e^{it\omega_k} \Big|,$$

and $(x_t - \mu)$ has mean zero and variance 1. Therefore even when $E(x_i) > 0$, (5.2) holds.

We still need to take care of the remaining eigenvalues λ_0 and $\lambda_{n/2}$ (if n is even), where

$$\lambda_0 = \frac{1}{\sqrt{n}} \sum_{t=0}^{n-1} x_t \quad \text{and} \quad \lambda_{n/2} = \frac{1}{\sqrt{n}} \sum_{t=0}^{n-1} (-1)^t x_t.$$

Note that by the Central Limit Theorem (CLT),

$$\frac{\sqrt{n}\lambda_0 - \mu n}{\sqrt{n}} \xrightarrow{\mathcal{D}} N(0,1). \tag{5.3}$$

(5.3) implies $\lambda_0 \xrightarrow{\mathcal{P}} \infty$ and also

$$|\lambda_0| - \mu\sqrt{n} \xrightarrow{\mathcal{D}} N(0,1).$$

Let $A_n = \max_{1 \leq k < q} |\lambda_k|$. From (5.2) and (5.3),

$$\frac{A_n}{\sqrt{\ln q}} \xrightarrow{\mathcal{P}} 1 \quad \text{and} \quad \frac{\lambda_0}{\mu\sqrt{n}} \xrightarrow{\mathcal{P}} 1,$$

and so it follows that

$$P\left[\max\{A_n, |\lambda_0|\} - \mu\sqrt{n} > x \right] \to P\left[N(0,1) > x \right].$$

This proves (a) for odd n when $\mu > 0$. Since for even n,

$$\lambda_{n/2} = \frac{1}{\sqrt{n}} \sum_{t=0}^{n-1} (-1)^t x_t \xrightarrow{\mathcal{D}} N(0,1),$$

this can also be neglected as before, and hence (a) holds also for even n. A similar proof works when $\mu < 0$. This proves (a) completely.

(b) Now assume $\mu = 0$. Then $|\lambda_0|$ is tight and

$$\frac{|\lambda_0| - \sqrt{\ln q}}{(\ln q)^{-1/2}} \xrightarrow{\mathcal{P}} -\infty.$$

Hence A_n dominates $|\lambda_0|$, and as a consequence, $\dfrac{\mathrm{sp}(\frac{1}{\sqrt{n}}RC_n) - d_q}{c_q} \xrightarrow{\mathcal{D}} \Lambda.$ $\square$

5.2   Symmetric circulant

The behavior of $\mathrm{sp}(\frac{1}{\sqrt{n}}SC_n)$ is similar to that of $\mathrm{sp}(\frac{1}{\sqrt{n}}RC_n)$ but the normalizing constants change. The following normalizing constants, well-known in the context of maxima of i.i.d. normal variables, will be used in the statements of our results.

$$a_n = (2\ln n)^{-1/2} \quad \text{and} \quad b_n = (2\ln n)^{1/2} - \frac{\ln\ln n + \ln 4\pi}{2(2\ln n)^{1/2}}. \tag{5.4}$$

The following lemmata are well known. The first lemma is on the joint behavior of the maxima and the minima of i.i.d. normal random variables. We leave its proof as an exercise. The second lemma is from Einmahl (1989) Corollary 1(b), page 31, in combination with his Remark on page 32. We omit its proof.

Lemma 5.2.1. Let $\{N_i\}$ be i.i.d. $N(0,1)$, $m_n = \min_{1 \le i \le n} N_i$ and $M_n = \max_{1 \le i \le n} N_i$. Then with a_n and b_n as in (5.4),

$$\left(\frac{-m_n - b_n}{a_n}, \frac{M_n - b_n}{a_n}\right) \xrightarrow{\mathcal{D}} \Lambda \otimes \Lambda,$$

where $\Lambda \otimes \Lambda$ is the joint distribution of two independent standard Gumbel random variables.

Let I_d be the $d \times d$ identity matrix and $|\cdot|$ be the Euclidean norm in $\mathbb{R}^d$. Let ϕ_C be the density of a d-dimensional centered normal random vector with covariance matrix C.

Lemma 5.2.2. Let $\{\psi_i\}$ be independent mean zero random vectors in $\mathbb{R}^d$ with finite moment generating functions in a neighborhood of the origin and $\mathrm{Cov}(\psi_1 + \psi_2 + \cdots + \psi_n) = B_n I_d$, where $B_n > 0$. Let $\{\eta_k\}$ be independent $N(0, \sigma^2 \mathrm{Cov}(\psi_k))$, independent of $\{\psi_k\}$, and $\sigma^2 \in (0,1]$. Let $\psi_k^* = \psi_k + \eta_k$, and let p_n^* denote the density of $B_n^{-1/2} \sum_{k=1}^n \psi_k^*$. Let $\alpha \in (0, \frac{1}{2})$ be such that $\alpha \sum_{k=1}^n \mathrm{E}\, |\psi_k|^3 \exp(\alpha|\psi_k|) \le B_n$. Let $\beta_n = \beta_n(\alpha) = B_n^{-3/2} \sum_{k=1}^n \mathrm{E}\, |\psi_k|^3 \exp(\alpha|\psi_k|)$. If $|x| \le c_1 \alpha B_n^{1/2}$, $\sigma^2 \ge -c_2 \beta_n^2 \ln \beta_n$ and $B_n \ge c_3 \alpha^{-2}$, where c_1, c_2, c_3 are constants depending only on d, then for some constant c_4 depending on d,

$$p_n^*(x) = \phi_{(1+\sigma^2)I_d}(x) \exp(\bar{T}_n(x)) \quad \text{with} \quad |\bar{T}_n(x)| \le c_4 \beta_n(|x|^3 + 1).$$

We shall use the above lemma now to derive a normal approximation result. This shall be used in the proof of Theorem 5.2.4 and then again in Section 8.3 of Chapter 8. For $n = 2j + 1$, define a triangular array of centered random variables $\{\bar{x}_t; 1 \leq t \leq 2j + 1\}$ by

$$\bar{x}_t = x_t \mathbb{I}(|x_t| \leq (1 + 2j)^{1/s}) - E[x_t \mathbb{I}(|x_t| \leq (1 + 2j)^{1/s})]. \tag{5.5}$$

For $1 \leq i_1 < i_2 < \cdots < i_d < j$ and $1 \leq t \leq j$, let

$$v_d(0) = \sqrt{2}(1, 1, \ldots, 1), \; v_d(t) = 2\left(\cos\frac{2\pi i_1 t}{2j + 1}, \cos\frac{2\pi i_2 t}{2j + 1}, \ldots, \cos\frac{2\pi i_d t}{2j + 1}\right).$$

Lemma 5.2.3. Let $n = 1 + 2j$ and $\sigma_j^2 = (1 + 2j)^{-c}$ for some $c > 0$ and let $\{x_t\}$ be i.i.d. mean zero with $E(x_0^2) = 1$ and $E|x_0|^s < \infty$ for some $s > 2$. Suppose N_t's are i.i.d. $N(0, 1)$ random variables independent of $\{x_t\}$ and $\tilde{p}_j(x)$ is the density of $\frac{1}{\sqrt{1+2j}}\sum_{t=0}^{j}(\bar{x}_t + \sigma_j N_t)v_d(t)$. Then for any measurable subset B of $\mathbb{R}^d$, there exists $\eta > 0$ and $\epsilon_j \to 0$ as $j \to \infty$, such that the following holds uniformly over d-tuples $1 \leq i_1 < i_2 < \cdots < i_d < j$:

$$\left|\int_B \tilde{p}_j(x)dx - \int_B \phi_{(1+\sigma_j^2)I_d}(x)dx\right| \leq \epsilon_j \int_B \phi_{(1+\sigma_j^2)I_d}(x)dx + O(\exp(-(1+2j)^\eta)).$$

Proof. Let $S_{j,\bar{x}} = \sum_{t=0}^{j}\bar{x}_t v_d(t)$ and let $s = 2 + \delta$. Then observe that $\text{Cov}(S_{j,\bar{x}}) = B_j I_d$, where $B_j = (2j + 1)\text{Var}(\bar{x}_t)$. Since $\{\bar{x}_t v_d(t)\}_{0 \leq t \leq j}$ are independent mean zero random vectors, we can use Lemma 5.2.2.

By choosing $\alpha = \frac{c_5(1+2j)^{-\frac{1}{s}}}{2\sqrt{d}}$, it easily follows that

$$\alpha \sum_{t=0}^{j} E|\bar{x}_t v_d(t)|^3 \exp(\alpha|x_t v_d(t)|) < B_j.$$

Let $\tilde{\beta}_j = B_j^{-3/2}\sum_{t=0}^{j} E|\bar{x}_t v_d(t)|^3 \exp(\alpha|\bar{x}_t v_d(t)|)$. Then clearly

$$\tilde{\beta}_j \leq C(1 + 2j)^{-(\frac{1}{2} - \frac{1-\delta}{s})}.$$

Let $c = \frac{1}{2} - \frac{1-\delta}{s} > 0$. Now choose $|x| \leq c_1\alpha B_j^{1/2} \approx c_2(1 + 2j)^{\frac{1}{2} - \frac{1}{s}}$ and σ_j^2 satisfying, $1 \geq \sigma_j^2 \geq c_3(\ln(2j + 1))(2j + 1)^{-2c}$. Clearly $B_j \geq c_4\alpha^{-2}$ and $B_j \approx (1 + 2j)$. Let c_1, c_2, c_3, c_4 be constants depending only on d. Then Lemma 5.2.2 implies that

$$\tilde{p}_j(x) = \phi_{(1+\sigma_j^2)I_d}(x)\exp(|T_j(x)|) \text{ with } |T_j(x)| \leq c_5\tilde{\beta}_j(|x|^3 + 1).$$

Note that $|T_j(x)| \to 0$ uniformly for $|x|^3 = o(\min\{(1 + 2j)^{-c}, (1 + 2j)^{\frac{1}{2} - \frac{1}{s}}\})$. For the choice of $\sigma_j^2 = (1 + 2j)^{-c}$, the above condition can be seen to be satisfied.

Now the following inequality can be shown for any measurable subset B of

$\mathbb{R}^d$. We omit its proof since a similar argument is provided in a more involved situation in Corollary 6.3.3 of Chapter 6.

$$\left| \int_B \tilde{p}_j(x)dx - \int_B \phi_{(1+\sigma_j^2)I_d}(x)dx \right| \le \epsilon_j \int_B \phi_{(1+\sigma_j^2)I_d}(x)dx + O(\exp(-(1+2j)^\eta)),$$

where $\epsilon_j \to 0$ as $j \to \infty$. This completes the proof. $\square$

For $\frac{1}{\sqrt{n}}RC_n$, leaving out the eigenvalues λ_0 and $\lambda_{n/2}$, due to perfect symmetry, the maximum and minimum eigenvalues are equal in magnitude. This is not the case for $\frac{1}{\sqrt{n}}SC_n$. Hence we should look at the joint behavior of the maximum and minimum eigenvalues.

Theorem 5.2.4 (Bose et al. (2011c)). Let $\{\lambda_k, 0 \le k \le n-1\}$ be the eigenvalues of $\frac{1}{\sqrt{n}}SC_n$. Let $q = \lfloor \frac{n}{2} \rfloor$, $M_{q,x} = \max_{1 \le k \le q} \lambda_k$ and $m_{q,x} = \min_{1 \le k \le q} \lambda_k$. If $\{x_i\}$ are i.i.d. with $\mathrm{E}(x_0) = 0$, $\mathrm{E}(x_0^2) = 1$, and $\mathrm{E}|x_0|^s < \infty$ for some $s > 2$, then
$$\left(\frac{-m_{q,x} - b_q}{a_q}, \frac{M_{q,x} - b_q}{a_q} \right) \xrightarrow{\mathcal{D}} \Lambda \otimes \Lambda,$$

where a_q and b_q are given by (5.4). The same limit continues to hold if the eigenvalue λ_0 is included in the definition of the max and the min above.

Proof. The proof is broken down into four steps. We use truncation and Lemma 5.2.3 along with Bonferroni inequality. First assume $n = 2j + 1$. Let $s = 2 + \delta$.

Step 1: Truncation. Let $\bar{x}_t$ be as in (5.5) and
$$\tilde{x}_t = x_t \mathbb{I}(|x_t| \le (1+2j)^{1/s}).$$

We show that it is enough to deal with the truncated random variables $\{\bar{x}_t\}$ (see (5.6) below). If $\bar{\lambda}_k$ and $\tilde{\lambda}_k$ denote the eigenvalues of $\frac{1}{\sqrt{n}}SC_n$ with input $\{\bar{x}_t\}$ and $\{\tilde{x}_t\}$ respectively, then $\bar{\lambda}_k = \tilde{\lambda}_k$. By Borel-Cantelli Lemma (see Section 11.2), $\sum_{t=1}^\infty |x_t|\mathbb{I}(|x_t| > t^{1/s})$ is bounded with probability 1 and consists of only a finite number of non-zero terms. Thus there exists a random positive integer N such that

$$\begin{aligned}
\sum_{t=0}^j |x_t - \tilde{x}_t| &= \sum_{t=0}^j |x_t|\mathbb{I}(|x_t| > (1+2j)^{1/s}) \\
&\le \sum_{t=0}^\infty |x_t|\mathbb{I}(|x_t| > t^{1/s}) \\
&= \sum_{t=0}^N |x_t|\mathbb{I}(|x_t| > t^{1/s}).
\end{aligned}$$

It follows that for $2j + 1 \geq \{N, |x_1|^s, \ldots, |x_N|^s\}$, the left side is zero. Consequently, for all j sufficiently large, $\bar{\lambda}_k = \lambda_k$ a.s. for all k. Therefore for all j sufficiently large,

$$\left(\frac{-m_{j,x} - b_j}{a_j}, \frac{M_{j,x} - b_j}{a_j}\right) \overset{\mathcal{D}}{=} \left(\frac{-m_{j,\bar{x}} - b_j}{a_j}, \frac{M_{j,\bar{x}} - b_j}{a_j}\right), \tag{5.6}$$

where

$$m_{j,\bar{x}} = \min_{1 \leq k \leq j} \bar{\lambda}_k \quad \text{and} \quad M_{j,\bar{x}} = \max_{1 \leq k \leq j} \bar{\lambda}_k.$$

Step 2: Application of Bonferroni Inequality. Define for $1 \leq k \leq j$,

$$\bar{\lambda}'_k = \frac{1}{\sqrt{2j+1}}\left(\sqrt{2}\bar{x}_0 + 2\sum_{t=1}^{j} \bar{x}_t \cos\frac{2\pi kt}{2j+1}\right),$$

$$\bar{\bar{\lambda}}'_k = \bar{\lambda}'_k + \frac{\sigma_j}{\sqrt{1+2j}}\left(\sqrt{2}N_0 + 2\sum_{t=1}^{j} N_t \cos\frac{2\pi kt}{n}\right)$$

$$= \bar{\lambda}'_k + \sigma_j N'_{j,k},$$

where $\sigma_j^2 = (1 + 2j)^{-c}$ for some $c > 0$. Observe that $N'_{j,k}$ are i.i.d. $N(0,1)$ for $k = 1, 2, \ldots, j$. Define

$$M_{j,\bar{x}+\sigma N} = \max_{1 \leq k \leq j} \bar{\bar{\lambda}}'_k \quad \text{and} \quad m_{j,\bar{x}+\sigma N} = \min_{1 \leq k \leq j} \bar{\bar{\lambda}}'_k.$$

Let

$$A = \left(\frac{-m_{j,\bar{x}+\sigma N} - b_j}{a_j} < x, \frac{M_{j,\bar{x}+\sigma N} - b_j}{a_j} < y\right) \text{ and}$$

$$B = \left(\frac{-\min_{1 \leq k \leq j}(1 + \sigma_j^2)N_k - b_j}{a_j} < x, \frac{\max_{1 \leq k \leq j}(1 + \sigma_j^2)N_k - b_j}{a_j} < y\right).$$

Claim:

$$\lim_{j \to \infty}[\mathrm{P}(A) - \mathrm{P}(B)] = 0. \tag{5.7}$$

Proof:

$$\mathrm{P}(A) = \mathrm{P}\left(\frac{-m_{j,\bar{x}+\sigma N} - b_j}{a_j} < x, \frac{M_{j,\bar{x}+\sigma N} - b_j}{a_j} < y\right)$$

$$= \mathrm{P}\left(m_{j,\bar{x}+\sigma N} > -a_j x - b_j, M_{j,\bar{x}+\sigma N} < a_j y + b_j\right)$$

$$= \mathrm{P}\left(\cap_{k=1}^{j}\{-(a_j x + b_j) < \bar{\bar{\lambda}}'_k < a_j y + b_j\}\right)$$

$$= 1 - \mathrm{P}\left(\cup_{k=1}^{j}\{\bar{\bar{\lambda}}'_k \in (I^j_{x,y})^c\}\right) = 1 - \mathrm{P}\left(\cup_{k=1}^{j} A_{k,j}\right),$$

where $I_{x,y}^j = (-a_j x - b_j, a_j y + b_j)$ and $A_{k,j} = \{\tilde{\lambda}_k' \in (I_{x,y}^j)^c\}$. By the Bonferroni inequality (see Section 11.2),

$$\sum_{t=1}^{2k}(-1)^{t-1}\tilde{A}_{t,j} \leq P(A) \leq \sum_{t=1}^{2k-1}(-1)^{t-1}\tilde{A}_{t,j}, \tag{5.8}$$

where

$$\tilde{A}_{t,j} = \sum_{1 \leq i_1 < i_2 < \cdots < i_t \leq j} P\left(A_{i_1,j} \cap \cdots \cap A_{i_t,j}\right).$$

$$P(B) = P\left(\frac{-\min_{1 \leq k \leq j}(1+\sigma_j^2)N_k - b_j}{a_j} < x, \frac{\max_{1 \leq k \leq j}(1+\sigma_j^2)N_k - b_j}{a_j} < y\right)$$

$$= 1 - P\left(\cup_{k=1}^j \{(1+\sigma_j^2)^{1/2}N_k \in (I_{x,y}^j)^c\}\right) = 1 - P\left(\cup_{k=1}^j B_{k,j}\right)$$

where $B_{k,j} = \{(1+\sigma_j^2)^{1/2}N_k \in (I_{x,y}^j)^c\}$. Again by the Bonferroni inequality,

$$\sum_{t=1}^{2k}(-1)^{t-1}\tilde{B}_{t,j} \leq P(B) \leq \sum_{t=1}^{2k-1}(-1)^{t-1}\tilde{B}_{t,j}, \tag{5.9}$$

where

$$\tilde{B}_{t,j} = \sum_{1 \leq i_1 < i_2 < \cdots < i_t \leq j} P\left(B_{i_1,j} \cap B_{i_2,j} \cap \cdots \cap B_{i_t,j}\right).$$

From (5.8) and (5.9), we get

$$\sum_{t=1}^{2k}(-1)^{t-1}(\tilde{A}_{t,j} - \tilde{B}_{t,j}) - \tilde{B}_{2k+1,j} \leq P(B) - P(A) \tag{5.10}$$

$$\leq \sum_{t=1}^{2k-1}(-1)^{t-1}(\tilde{A}_{t,j} - \tilde{B}_{t,j}) + \tilde{B}_{2k,j}.$$

Now note that

$$\tilde{B}_{t,j} = \sum_{1 \leq i_1 < i_2 \cdots < i_t \leq j} P\left((1+\sigma_j^2)^{1/2}N_{i_l} \in (I_{x,y}^j)^c; l = 1, 2, \ldots, t\right)$$

$$= \sum_{1 \leq i_1 < i_2 \cdots < i_t \leq j} [P\left((1+\sigma_j^2)^{1/2}N_{i_l} \in (I_{x,y}^j)^c\right)]^t.$$

Observe that

$$P\left((1+\sigma_j^2)^{1/2}N_1 \in (-a_j x - b_j, a_j y + b_j)^c\right)$$

$$\leq P\left((1+\sigma_j^2)^{1/2}N_1 < -a_j x - b_j\right) + P\left((1+\sigma_j^2)^{1/2}N_1 > a_j y + b_j\right)$$

$$= P\left(N_1 < -(a_j x + b_j)(1+\sigma_j^2)^{-1/2}\right) + P\left(N_1 > (a_j y + b_j)(1+\sigma_j^2)^{-1/2}\right)$$

$$\leq P\left(N_1 < -(a_j x + b_j)(1 - \frac{1}{2}\sigma_j^2)\right) + P\left(N_1 > (a_j y + b_j)(1 - \frac{1}{2}\sigma_j^2)\right).$$

Now $(a_j y + b_j)(1 - \frac{\sigma_j^2}{2}) \approx b_j + o(1)$ and $P(N_1 > b_j) \approx \frac{1}{j}$. We can estimate the left tail similarly. Therefore

$$P\left(N_1 < -(a_j x + b_j)(1 - \frac{1}{2}\sigma_j^2)\right) + P\left(N_1 > (1 - \frac{1}{2}\sigma_j^2)(a_j y + b_j)\right) \leq \frac{C}{j},$$

and hence

$$\lim_{t \to \infty} \overline{\lim_{j \to \infty}} \tilde{B}_{t,j} \leq \lim_{t \to \infty} \overline{\lim_{j \to \infty}} \binom{j}{t} \frac{C^t}{j^t} \leq \lim_{t \to \infty} \frac{C^t}{t!} = 0.$$

On the other hand, fixing $t \geq 1$ we get

$$P(A_{i_1,j} \cap A_{i_2,j} \cap \cdots \cap A_{i_t,j}) = P(\frac{1}{\sqrt{1+2j}} \sum_{t=0}^{j} (\bar{x}_t + \sigma_j N_t) v_d(t) \in E_t),$$

where $E_t = \{(x_1, x_2, \ldots, x_t) : x_i \in (I_{x,y}^j)^c\}$.

So by Lemma 5.2.3 we have that uniformly over all d-tuples $1 \leq i_1 < i_2 < \cdots < i_d \leq j$,

$$\left| P(\frac{1}{\sqrt{1+2j}} \sum_{t=0}^{j} (\bar{x}_t + \sigma_j N_t) v_d(t) \in E_t) - P((1+\sigma_j^2)^{1/2} N_{i_l} \in (I_{x,y}^j)^c, 1 \leq l \leq t) \right|$$

$$\leq \epsilon_j P((1+\sigma_j^2)^{1/2} N_{i_l} \in (I_{x,y}^j)^c, 1 \leq l \leq t) + O(\exp(-(1+2j)^\eta)).$$

So for fixed t, as $j \to \infty$, we get

$$|\tilde{A}_{t,j} - \tilde{B}_{t,j}| \leq \epsilon_j \tilde{B}_{t,j} + \binom{j}{t} O(\exp(-(1+2j)^\eta)) \to 0.$$

Therefore

$$\overline{\lim_{j \to \infty}} |P(A) - P(B)| \leq \overline{\lim_{j \to \infty}} \tilde{B}_{2k+1,j} + \overline{\lim_{j \to \infty}} \tilde{B}_{2k,j} \to 0 \text{ as } k \to \infty.$$

This proves the claim (5.7) completely.

Step 3: Claim:

$$\left(\frac{-m_{j,\bar{x}+\sigma N} - b_j}{a_j}, \frac{M_{j,\bar{x}+\sigma N} - b_j}{a_j}\right) \xrightarrow{\mathcal{D}} \Lambda \otimes \Lambda. \tag{5.11}$$

As $\max_{1 \leq k \leq j} N_k = O_p((\ln j)^{1/2})$, it follows that

$$\left| \frac{(1+\sigma_j^2)^{1/2} \max_{1 \leq k \leq j} N_k - b_j}{a_j} - \frac{\max_{1 \leq k \leq j} N_k - b_j}{a_j} \right| \leq \frac{\sigma_j \max_{1 \leq k \leq j} |N_k|}{a_j}$$

$$\xrightarrow{\mathcal{P}} 0.$$

Therefore

$$\frac{(1+\sigma_j^2)^{1/2}\max_{1\le k\le j} N_k - b_j}{a_j} \xrightarrow{\mathcal{D}} \Lambda.$$

Since $-\min_{1\le k\le j}(1+\sigma_j^2)^{1/2}N_k = \max_{1\le k\le j}\left(-(1+\sigma_j^2)^{1/2}N_k\right)$ and $-(1+\sigma_j^2)^{1/2}N_k \overset{\mathcal{D}}{=} (1+\sigma_j^2)^{1/2}N_k$, we get

$$\frac{\min_{1\le k\le j} -(1+\sigma_j^2)^{1/2}N_k - b_j}{a_j} \xrightarrow{\mathcal{D}} \Lambda.$$

Since $(1+\sigma_j^2)^{1/2}N_i$ are i.i.d. symmetric random variables, by Resnick (1987) (Exercise 5.5.2),

$$\left(\frac{\min_{1\le k\le j} -(1+\sigma_j^2)^{1/2}N_k - b_j}{a_j}, \frac{\max_{1\le k\le j}(1+\sigma_j^2)^{1/2}N_k - b_j}{a_j}\right) \xrightarrow{\mathcal{D}} \Lambda \otimes \Lambda.$$

$$(5.12)$$

Therefore combining (5.7) and (5.12), we get

$$\left(\frac{-m_{j,\bar{x}+\sigma N} - b_j}{a_j}, \frac{M_{j,\bar{x}+\sigma N} - b_j}{a_j}\right) \xrightarrow{\mathcal{D}} \Lambda \otimes \Lambda.$$

This completes the proof of (5.11).

Step 4: Claim:

$$\left(\frac{-m_{j,\bar{x}} - b_j}{a_j}, \frac{M_{j,\bar{x}} - b_j}{a_j}\right) \xrightarrow{\mathcal{D}} \Lambda \otimes \Lambda. \qquad (5.13)$$

We prove this using (5.11). Note that

$$\left|\frac{\max(\bar{\bar{\lambda}}_k')}{a_j} - \frac{\max(\bar{\lambda}_k')}{a_j}\right| \le \frac{\sigma_j}{a_j}\max|N_{j,k}'| \xrightarrow{\mathcal{P}} 0.$$

Similarly $-\bar{\bar{\lambda}}_k' = -\bar{\lambda}_k' - \sigma N_{j,k}'$, and

$$\left|\frac{\max(-\bar{\bar{\lambda}}_k')}{a_j} - \frac{\max(-\bar{\lambda}_k')}{a_j}\right| \le \frac{\sigma_j}{a_j}\max|N_{j,k}'| \xrightarrow{\mathcal{P}} 0.$$

Now if we denote $m_{j,\bar{x}}' = \min_{1\le k\le j}\bar{\lambda}_k'$ and $M_{j,\bar{x}}' = \max_{1\le k\le j}\bar{\lambda}_k'$, then

$$\left|\left(\frac{-m_{j,\bar{x}+\sigma N} - b_j}{a_j}, \frac{M_{j,\bar{x}+\sigma N} - b_j}{a_j}\right) - \left(\frac{-m_{j,\bar{x}}' - b_j}{a_j}, \frac{M_{j,\bar{x}}' - b_j}{a_j}\right)\right|$$

$$\le C\left[\left|\frac{-m_{j,\bar{x}+\sigma N} - (-m_{j,\bar{x}}')}{a_j}\right| + \left|\frac{M_{j,\bar{x}+\sigma N} - M_{j,\bar{x}}'}{a_j}\right|\right]$$

$$\le C\left[\left|\frac{\max(-\bar{\bar{\lambda}}_k') - \max(-\bar{\lambda}_k')}{a_j}\right| + \left|\frac{\max(\bar{\bar{\lambda}}_k') - \max(\bar{\lambda}_k')}{a_j}\right|\right] \xrightarrow{\mathcal{P}} 0.$$

Therefore using (5.11), we have

$$\left(\frac{-m'_{j,\bar{x}} - b_j}{a_j}, \frac{M'_{j,\bar{x}} - b_j}{a_j}\right) \xrightarrow{\mathcal{D}} \Lambda \otimes \Lambda. \tag{5.14}$$

Again $\bar{\lambda}_k = \bar{\lambda}'_k + \frac{(1-\sqrt{2})}{\sqrt{2j+1}}\bar{x}_0$. Therefore

$$\left|\frac{M'_{j,\bar{x}} - b_j}{a_j} - \frac{M_{j,\bar{x}} - b_j}{a_j}\right| + \left|\frac{-m_{j,\bar{x}} - b_j}{a_j} - \frac{-m'_{j,\bar{x}} - b_j}{a_j}\right| \xrightarrow{\mathcal{P}} 0.$$

Hence using (5.14), we have

$$\left(\frac{-m_{j,\bar{x}} - b_j}{a_j}, \frac{M_{j,\bar{x}} - b_j}{a_j}\right) \xrightarrow{\mathcal{D}} \Lambda \otimes \Lambda.$$

This completes the proof of (5.13).

Now we get back to the main proof. Combining (5.6) and (5.13),

$$\left(\frac{-m_{j,x} - b_j}{a_j}, \frac{M_{j,x} - b_j}{a_j}\right) \xrightarrow{\mathcal{D}} \Lambda \otimes \Lambda.$$

This proves the theorem when n is odd. Now suppose $n = 2j$. If we work with

$$\lambda'_k = \sqrt{2}x_0 + \sqrt{2}(-1)^k x_j + 2\sum_{t=1}^{j-1} x_t \cos\frac{2\pi kt}{2j},$$

then similar normal approximations can be done and subsequent calculations follow. We omit the obvious details. This proves the theorem completely. $\square$

Theorem 5.2.4 implies that the distribution of the range of the spectrum is the convolution of two Gumbel distributions. So one can ask a very natural question: what happens in general to the spectral gaps? We shall address this question in Chapter 8.

The next theorem is similar to Theorem 5.1.1. We leave the proof as an exercise.

Theorem 5.2.5 (Bose et al. (2011c)). Suppose the input $\{x_i\}_{i\geq 0}$ is i.i.d. with mean μ, variance one, and $E|x_i|^{2+\delta} < \infty$ for some $\delta > 0$.

(a) If $\mu = 0$ then

$$\frac{\text{sp}(\frac{1}{\sqrt{n}}SC_n) - b_q - a_q \ln 2}{a_q} \xrightarrow{\mathcal{D}} \Lambda,$$

where $q = q(n) \approx \frac{n}{2}$ and a_q, b_q are as in equation (5.4).

(b) If $\mu \neq 0$ then

$$\frac{\text{sp}(SC_n) - |\mu|n}{\sqrt{n}} \xrightarrow{\mathcal{D}} N(0, 2).$$

5.3 Exercises

1. Prove Lemma 5.2.1.

2. Prove Theorem 5.2.5.

3. For any matrix A with possibly complex entries, its *spectral norm* $\|A\|$ is defined as the square root of the largest eigenvalue of the positive semi-definite matrix A^*A:

$$\|A\| = \sqrt{\lambda_{max}(A^*A)},$$

   where A^* denotes the conjugate transpose of A.

   Show that if A is an $n \times n$ real symmetric matrix or A is a normal matrix (that is, $A^*A = AA^*$), then

$$\|A\| = \mathtt{sp}(A).$$

4. Find the limiting distribution of the spectral norm of C_n, RC_n and SC_n using proper scaling and centering. Hint: Note that RC_n, SC_n are symmetric and C_n is a normal matrix.

5. Consider RC_n with input $\{x_i\}$ which is i.i.d. with $\mathrm{E}(x_0) = \mu$ and $\mathrm{Var}(x_0) = 1$. Let $RC_n^0 = RC_n - \mu n u_n u_n^T$ where $u_n = \frac{1}{\sqrt{n}}(1, 1, \dots, 1)^T$. If $\mu > 0$, then show that

$$\frac{\mathtt{sp}(RC_n)}{n} \to \mu \text{ almost surely, and } \left\| \frac{RC_n^0}{\|RC_n\|} \right\| \to 0 \text{ almost surely.}$$

   Hint: For the second part see Lemma 1(i) and Theorem 3 of Bose and Sen (2007).

6. Look up the proof of Lemma 5.2.2.

7. Look up Exercise 5.5.2 from Resnick (1987).

8. Look up the results on the behavior of the maximum of the singular values of C_n with standard complex normal entries from Meckes (2009).

6

Spectral radius: k-circulant

This chapter is devoted to extending the results of the previous chapter to the k-circulant when $n = k^g + 1$. The main result says that the centered and the scaled spectral radius converges to the Gumbel distribution. To establish this result, we make some intricate study of the eigenvalues of the matrix. We also need an auxiliary result which exhibits the tail behavior of the n-fold product of i.i.d. standard exponential variables. Finally, we also need truncation and normal approximation similar to the ones we used in the previous chapter.

6.1 Tail of product

Let $\{E_i\}$ be i.i.d. standard exponentials. Define

$$H_n(x) = P[E_1 E_2 \cdots E_n > x]. \tag{6.1}$$

Its behavior as $x \to \infty$ is outlined in the next theorem. It implies that H_n lies in the maximum domain of attraction of the Gumbel distribution for any fixed n.

Theorem 6.1.1 (Bose et al. (2012a)).

$$H_n(x) = C_n x^{\alpha_n} e^{-n x^{\frac{1}{n}}} g_n(x), \quad n \geq 1, \tag{6.2}$$

where for $n \geq 1$,

$$C_n = \frac{1}{\sqrt{n}} (2\pi)^{\frac{n-1}{2}}, \quad \alpha_n = \frac{n-1}{2n}, \quad \text{and} \quad g_n(x) \to 1 \quad \text{as} \quad x \to \infty.$$

Proof of Theorem 6.1.1. We shall use the method of induction. Since $H_1(x) = P[E_1 > x] = e^{-x}$, we can choose $C_1 = 1$, $\alpha_1 = 0$ and $g_1(x) = 1$ for all x. Hence the result is true for $n = 1$. For larger n, we need *Laplace's asymptotics.* We give a brief description of this for our purposes. For more details, see Section 2.4 of Erdélyi (1956). Let

$$I(x) = \int_a^b f(t) e^{-xg(t)} dt, \quad x > 0,$$

where $f(\cdot)$ is continuous, $g(\cdot)$ is twice continuously differentiable, and $g(\cdot)$ reaches its only minimum over $[a,b]$ at an interior point c. Then

$$I(x) = e^{-xg(c)}f(c)\sqrt{\frac{2\pi}{xg''(c)}}(1+o(1)), \quad \text{as } x \to \infty. \tag{6.3}$$

Now

$$
\begin{aligned}
H_2(x) &= \int_0^\infty e^{-y}e^{-x/y}dy \\
&= x^{1/2}\int_0^\infty e^{-x^{1/2}(t+\frac{1}{t})}dt \quad \text{(substituting } y = tx^{1/2}) \\
&= x^{1/2}\int_0^\infty f(t)e^{-x^{1/2}g(t)}dt,
\end{aligned}
$$

where $f(t) = 1$ and $g(t) = t + \frac{1}{t}$. Note that g assumes a strict minimum at $t = 1$ and $f(1) = 1$. Hence, applying (6.3) we have

$$
\begin{aligned}
H_2(x) &= x^{1/2}e^{-x^{1/2}g(1)}f(1)\sqrt{\frac{2\pi}{x^{1/2}g''(1)}}g_2(x) \\
&= \sqrt{\pi}x^{1/4}e^{-2x^{1/2}}g_2(x),
\end{aligned}
$$

where $g_2(x) \to 1$ as $x \to \infty$. Hence, $C_2 = \sqrt{\pi} = \frac{1}{\sqrt{2}}(2\pi)^{1/2}$ and $\alpha_k = 1$. So the result is true for $n = 2$. Now suppose (6.2) is true for $n = k$. We shall prove it for $n = k+1$.

$$
\begin{aligned}
H_{k+1}(x) &= \int_0^\infty e^{-y}H_k(\frac{x}{y})dy \\
&= C_k\int_0^\infty e^{-y}(\frac{x}{y})^{\alpha_k}e^{-k(\frac{x}{y})^{1/k}}g_k(\frac{x}{y})dy \\
&= xkC_k\int_0^\infty e^{-(ks+\frac{x}{s^k})}s^{k\alpha_k-k-1}g_k(s^k)ds \quad (x/y = s^k) \\
&= x^{\frac{k\alpha_k+1}{k+1}}kC_k\int_0^\infty e^{-(kt+\frac{1}{t^k})x^{\frac{1}{k+1}}}t^{k\alpha_k-k-1}g_k(t^kx^{\frac{k}{k+1}})dt \quad (s = x^{\frac{1}{k+1}}t) \\
&= x^{\frac{k\alpha_k+1}{k+1}}kC_k\int_0^\infty f(t)e^{-x^{\frac{1}{k+1}}g(t)}dt,
\end{aligned}
$$

where

$$f(t) = t^{k\alpha_k-k-1}g_k(t^kx^{\frac{k}{k+1}}) \quad \text{and} \quad g(t) = kt + \frac{1}{t^k}.$$

Note that g assumes a strict minimum at $t = 1$. Moreover,

$$f(1) = g_k(x^{\frac{k}{k+1}}) \neq 0, \quad \text{and} \quad g''(1) = k(k+1).$$

Hence, applying (6.3) we have

$$
\begin{aligned}
H_{k+1}(x) &= x^{\frac{k\alpha_{k+1}}{k+1}} kC_k e^{-x^{\frac{1}{k+1}}g(1)} f(1)\sqrt{\frac{2\pi}{x^{\frac{1}{k+1}}g''(1)}} h(x) \\
&= x^{\frac{k\alpha_{k+1}}{k+1}} kC_k e^{-(k+1)x^{\frac{1}{k+1}}} g_k(x^{\frac{k}{k+1}})\sqrt{\frac{2\pi}{x^{\frac{1}{k+1}}k(k+1)}} h(x),
\end{aligned}
$$

where $h(x) \to 1$ as $x \to \infty$. Substituting the values of α_k and C_k we get

$$
\begin{aligned}
H_{k+1}(x) &= x^{\frac{k}{2(k+1)}} \frac{1}{\sqrt{k+1}} (2\pi)^{k/2} e^{-(k+1)x^{\frac{1}{k+1}}} g_k(x^{\frac{k}{k+1}})h(x) \\
&= C_{k+1}x^{\alpha_{k+1}} e^{-(k+1)x^{\frac{1}{k+1}}} g_{k+1}(x),
\end{aligned}
$$

where

$$
\alpha_{k+1} = \frac{k}{2(k+1)}, \quad C_{k+1} = \frac{1}{\sqrt{k+1}}(2\pi)^{k/2}, \quad g_{k+1}(x) = g_k(x^{\frac{k}{k+1}})h(x),
$$

and $\lim_{x\to\infty} g_{k+1}(x) = 1$. So the result is true for $n = k+1$ and the proof is complete. $\qquad\square$

The next theorem follows easily from the standard calculations in extreme value theory as found in Resnick (1987) and Embrechts et al. (1997). For the sake of completeness we give a proof here.

Theorem 6.1.2 (Bose et al. (2012a)). Let $\{X_n\}$ be a sequence of i.i.d. non-negative random variables with distribution F and let $F^{(n)} = \max_{1\le i\le n} X_i$. If $1 - F(x) \sim Cx^b e^{-ax^2}$ as $x \to \infty$, then

$$
\frac{F^{(n)} - d_n}{c_n} \xrightarrow{\ D\ } \Lambda,
$$

where

$$
c_n = \frac{1}{2a^{1/2}(\ln n)^{1/2}} \quad \text{and}
$$

$$
d_n = \frac{\ln C - \frac{b}{2}\ln a}{2a^{1/2}(\ln n)^{1/2}} + \Big(\frac{\ln n}{a}\Big)^{1/2}\Big[1 + \frac{b\ln\ln n}{4\ln n}\Big].
$$

Proof. Let $\overline{F} = 1 - F$. Then

$$
\begin{aligned}
\overline{F}(x) &= \theta(x)\overline{F}_\#(x), \quad \text{where} \\
\theta(x) &\to \theta = Ce^{-a} \quad \text{and} \quad \overline{F}_\#(x) = x^b\exp(-a(x^2 - 1)).
\end{aligned} \tag{6.4}
$$

By invoking Proposition 1.1 of Resnick (1987) (Proposition 11.2.1 in Appendix), it is enough to show the following: there exists some x_0 and a function

f such that $f(y) > 0$ for $y > x_0$, and such that f has an absolute continuous derivative with $f'(x) \to 0$ as $x \to \infty$ so that,

$$\overline{F}_\#(x) = 1 - F_\#(x) = \exp\left(-\int_{x_0}^{x}(1/f(y))dy\right), \quad x > x_0. \tag{6.5}$$

Further, a choice for the normalizing constants (for $\overline{F}_\#(x)$) is given by

$$d_n^* = (1/(\overline{F}_\#))^{-1}(n) \quad \text{and} \quad c_n^* = f(d_n^*). \tag{6.6}$$

Comparing the two representations of $\overline{F}_\#$ given in (6.4) and (6.5), the following choice of f verifies (6.5):

$$f(x) = \frac{x}{2ax^2 - b} \sim \frac{1}{2ax} \quad \text{as } x \to \infty.$$

Now we identify d_n^* and c_n^*. Noting that $d_n^* \to \infty$ and $\overline{F}_\#(d_n^*) = \frac{1}{n}$, we get

$$c_n^* = f(d_n^*) \sim \frac{1}{2ad_n^*} \quad \text{and} \quad (d_n^*)^b \exp(-a((d_n^*)^2 - 1)) = \frac{1}{n}.$$

Taking logarithms on both sides of the second equation, we have

$$a d_n^{*\,2} - b \ln d_n^* - a = \ln n. \tag{6.7}$$

Let $d_n^* = \left(\frac{\ln n}{a}\right)^{1/2}(1 + \delta_n)$. Using this in (6.7) we get

$$\delta_n = \frac{\frac{b}{2}\ln\ln n + \epsilon_n}{2\ln n} + O\left(\frac{(\ln\ln n)^2}{(\ln n)^2}\right),$$

where $\epsilon_n = -b\ln(1 + \delta_n) - \frac{b}{2}\ln a + a$. So we get

$$
\begin{aligned}
d_n^* &= \left(\frac{\ln n}{a}\right)^{1/2}\left[1 + \frac{b}{4}\frac{\ln\ln n}{\ln n} + \frac{a - \frac{b}{2}\ln a - b\ln(1 + \delta_n)}{2\ln n}\right] + O\left(\frac{(\ln\ln n)^2}{(\ln n)^{3/2}}\right) \\
&\sim \left(\frac{\ln n}{a}\right)^{1/2}\left[1 + \frac{b}{4}\frac{\ln\ln n}{\ln n} + \frac{a - \frac{b}{2}\ln a}{2\ln n}\right] = \hat{d}_n \text{ (say), and} \\
c_n &= \frac{1}{2a^{1/2}(\ln n)^{1/2}}.
\end{aligned}
$$

Hence

$$\frac{F^{(n)} - \hat{d}_n}{c_n} \xrightarrow{\mathcal{D}} \Lambda_{Ce^{-a}}.$$

The result follows by using $d_n = c_n \ln(Ce^{-a}) + \hat{d}_n$ and convergence of types result (Proposition 0.2 of Resnick (1987), see Section 11.2 in Appendix). $\quad\square$

The following two corollaries follow immediately from Theorems 6.1.1 and 6.1.2. They will be useful in the proofs of Theorem 6.4.1 and Lemma 6.4.2, respectively.

Corollary 6.1.3. Let $\{X_n\}$ be a sequence of i.i.d. random variables where $X_i \overset{\mathcal{D}}{=} (E_1 E_2 \cdots E_k)^{1/2k}$ and $\{E_i\}$ are i.i.d. $\text{Exp}(1)$ random variables. Then

$$\frac{\max_{1 \le i \le n} X_i - d_n}{c_n} \overset{\mathcal{D}}{\to} \Lambda,$$

where Λ is the standard Gumbel distribution,

$$c_n = \frac{1}{2k^{1/2}(\ln n)^{1/2}},$$

$$d_n = \frac{\ln C_k - \frac{k-1}{2} \ln k}{2k^{1/2}(\ln n)^{1/2}} + \left(\frac{\ln n}{k}\right)^{1/2}\left[1 + \frac{(k-1)\ln \ln n}{4 \ln n}\right], \text{ and}$$

$$C_k = \frac{1}{\sqrt{k}}(2\pi)^{\frac{k-1}{2}}.$$

Corollary 6.1.4. Let $\{E_i\}$, c_n and d_n be as in Corollary 6.1.3. Let $\sigma_n^2 = n^{-c}$, $c > 0$. Then there exists some positive constant $K = K(x)$, such that for all large n we have

$$\mathrm{P}\left((E_1 E_2 \cdots E_k)^{1/2k} > (1 + \sigma_n^2)^{-1/2}(c_n x + d_n)\right) \le \frac{K}{n}, \quad x \in \mathbb{R}.$$

6.2 Additional properties of the k-circulant

Before we study the extreme of the k-circulant, we need some additional facts about their eigenvalues. Recall the eigenvalue structure of the k-circulant matrices from Section 1.4 of Chapter 1. Also recall that $S(x) = \{xk^b \pmod{n'} : 0 \le b < g_x\}$ and $g_x = \#S(x)$, the *order* of x. Note that $g_0 = 1$. Recall from (1.8) and (1.10) of Chapter 1 that

$$\mathbb{Z}_n = \{0, 1, 2, \ldots, n-1\},$$
$$y_j = \prod_{t \in \mathcal{P}_j} \lambda_{ty}, \quad j = 0, 1, \ldots, l-1 \quad \text{where} \quad y = n/n', \quad \text{and}$$
$$\upsilon_{k,n'} = \#\{x \in \mathbb{Z}_{n'} : g_x < g_1\}.$$

Here we focus on the k-circulant matrix with $n = k^g + 1$ and observe that $n = n'$. Define

$$J_k := \{\mathcal{P}_i : \#\mathcal{P}_i = k\}, \quad n_k := \#J_k, \text{ and}$$
$$X(k) := \{x : x \in \mathbb{Z}_n \text{ and } x \text{ has order } k\}.$$

Lemma 6.2.1. The eigenvalues $\{\eta_i\}$ of the k-circulant with $n = k^g + 1$, $g \ge 2$, satisfy the following:

(a) $\eta_0 = \sum_{t=0}^{n-1} x_t$, is always an eigenvalue and if n is even, then $\eta_{\frac{n}{2}} = \sum_{t=0}^{n-1}(-1)^t x_t$, is also an eigenvalue and both have multiplicity one.

(b) For $x \in \mathbb{Z}_n \setminus \{0, \frac{n}{2}\}$, $g_x = g_1$ or $\frac{g_1}{b}$ for some $b \geq 2$ and $\frac{g_1}{2b}$ is an integer.

(c) For all large n, $g_1 = 2g$. Hence from (b), for $x \in \mathbb{Z}_n \setminus \{0, \frac{n}{2}\}$, $g_x = 2g$ or $\frac{2g}{b}$. The total number of eigenvalues corresponding to J_{2g} is

$$2g \times \#J_{2g} = \#X(2g) \approx n.$$

(d) $X(\frac{2g}{b}) = \emptyset$ for $2 \leq b < g$, b even. If g is even then $X(\frac{2g}{g}) = X(2)$ is either empty or contains exactly two elements with eigenvalues

$$\eta_l = |\lambda_l| \quad \text{and} \quad \eta_{n-l} = -|\lambda_l|, \quad \text{for some } 1 \leq l \leq \frac{n}{2}.$$

(e) Suppose b is odd, $3 \leq b \leq g$ and $\frac{g}{b}$ is an integer. For each $\mathcal{P}_j \in J_{\frac{2g}{b}}$, there are $\frac{2g}{b}$ eigenvalues given by the $(\frac{2g}{b})$-th roots of y_j. The total number of eigenvalues corresponding to the set $J_{\frac{2g}{b}}$ is

$$\frac{2g}{b} \times \#J_{\frac{2g}{b}} = \#X(\frac{2g}{b}) \approx (k^{g/b}+1)(1+n^{-a}) \quad \text{for some } a > 0.$$

There are no other eigenvalues.

Proof. Since n and k are relatively prime, we have $n' = n$.

(a) $\mathcal{P}_0 = S(0) = \{0\}$ and the corresponding eigenvalue is $\eta_0 = \sum_{t=0}^{n-1} x_t$ with multiplicity one. If n is even then k is odd and hence $S(n/2) = \{\frac{n}{2}\}$, and the corresponding eigenvalue is $\eta_{\frac{n}{2}} = \sum_{t=0}^{n-1}(-1)^t x_t$ of multiplicity one.

(b) From Lemma 1.4.2(b), g_x divides g_1 and hence $g_x = g_1$ or $g_x = \frac{g_1}{b}$ for some $b \geq 2$. Also for every integer $t \geq 0$, $tk^g = (-1+n)t = -t \pmod{n}$. Hence, λ_t and λ_{n-t} belong to the same partition block $S(t) = S(n-t)$. Thus, each $S(t)$ contains an even number of elements, except for $t = 0, \frac{n}{2}$. Hence $\frac{g_1}{b}$ must be even, that is, $\frac{g_1}{2b}$ must be an integer.

(c) From Lemma 4.5.2(a), $g_1 = 2g$ for all but finitely many n, and $v_{k,n}/n \to 0$ as $n \to \infty$. For each $\mathcal{P}_j \in J_{2g}$, we have $2g$-many eigenvalues which are the $2g$-th roots of y_j. Now the result follows from the fact that $n = 2g\#J_{2g} + v_{k,n}$.

(d) Suppose $b = 2$ and $x \in X(\frac{g_1}{2}) = X(\frac{2g}{2})$. Then $xk^{\frac{g_1}{2}} = xk^g = x \pmod{n}$. But $k^g = -1 \pmod{n}$ and so, $xk^g = -x \pmod{n}$. Therefore $2x = 0 \pmod{n}$, and hence x is either 0 or $n/2$. But from (a), $g_0 = g_{n/2} = 1$. Hence $X(\frac{2g}{2}) = \emptyset$.

Now suppose $b > 2$ and is even. From Lemma 1.4.4(b),

$$\#X(\frac{2g}{b}) \leq \gcd(k^{2g/b} - 1, k^g + 1) \quad \text{for } b \geq 3.$$

Now observe that for b even,

$$\gcd(k^{2g/b} - 1, k^g + 1) = \begin{cases} 1 & \text{if } k \text{ is even,} \\ 2 & \text{if } k \text{ is odd.} \end{cases}$$

So we have $\#X(\frac{2g}{b}) \leq 2$ for $b > 2$ and b even.

Suppose if possible, there exists an $x \in \mathbb{Z}_n$ such that $g_x = \frac{2g}{b}$. Then $\#S(x) = \frac{2g}{b}$ and for all $y \in S(x)$, $g_y = \frac{2g}{b}$. Hence

$$\#\{y : g_y = \frac{2g}{b}\} \geq \frac{2g}{b} > 2 \quad \text{for} \quad g > b > 2, \ b \text{ even.}$$

This contradicts the fact that $\#X(\frac{2g}{b}) \leq 2$ for $g > b > 2$, b even. Hence $X(\frac{2g}{b}) = \emptyset$ for b even and $g > b > 2$.

If $b = g$ and is even, then from the previous discussion $\#X(\frac{2g}{g}) = 0$ or 2. In the latter case there are exactly two elements in $\mathbb{Z}_n$ whose orders are 2 and there will be only one partitioning set containing them. So the corresponding eigenvalues will be $\eta_l = |\lambda_l|$ and $\eta_{n-l} = -|\lambda_l|$, for some $1 \leq l \leq \frac{n}{2}$.

(e) We first show the following for b odd. Note that (e) is a simple consequence of this.

$$(k^{g/b} + 1) - \sum_*(k^{g/b_i} + 1) \leq \#X(\frac{2g}{b}) \leq k^{g/b} + 1, \tag{6.8}$$

where $\sum_*$ is sum over all odd $b_i > b$, such that $\frac{g}{b_i}$ is a positive integer. Let

$$Z_{n,b} = \{x : x \in \mathbb{Z}_n \text{ and } xk^{2g/b} = x \mod (k^g + 1)\}. \tag{6.9}$$

Then it is easy to see that

$$X(\frac{2g}{b}) \subseteq Z_{n,b}. \tag{6.10}$$

Let $x \in Z_{n,b}$ and $\frac{g}{b} = m$. Then

$$k^g + 1 \mid x(k^{2g/b} - 1)$$
$$\Rightarrow \quad k^{bm} + 1 \mid x(k^{2m} - 1)$$
$$\Rightarrow \quad k^{(b-1)m} - k^{(b-2)m} + k^{(b-3)m} - \cdots - k + 1 \mid x(k^m - 1).$$

But $\gcd(k^m - 1, k^{(b-1)m} - k^{(b-2)m} + k^{(b-3)m} - \cdots - k + 1) = 1$, and therefore x is a multiple of $(k^{(b-1)m} - k^{(b-2)m} + k^{(b-3)m} - \cdots - k + 1)$. Hence

$$\begin{aligned} \#Z_{n,b} &= \left\lfloor \frac{k^{bm} + 1}{(k^{(b-1)m} - k^{(b-2)m} + k^{(b-3)m} - \cdots - k + 1)} \right\rfloor \\ &= k^m + 1 \\ &= k^{g/b} + 1, \end{aligned}$$

and combining with (6.10),

$$\#X(\frac{2g}{b}) \leq \#Z_{n,b} = k^{g/b} + 1,$$

establishing the right side inequality in (6.8). On the other hand, if $x \in Z_{n,b}$

then either $g_x = \frac{2g}{b}$ or $g_x < \frac{2g}{b}$. For the second case $g_x = \frac{2g}{b_i}$ for some $b_i > b$, b_i odd, and therefore $x \in Z_{n,b_i}$. Hence

$$
\begin{aligned}
\#X\Big(\frac{2g}{b}\Big) &\geq \#Z_{n,b} - \sum_{*} \#Z_{n,b_i} \\
&\geq (k^{g/b} + 1) - \sum_{*}(k^{g/b_i} + 1),
\end{aligned}
$$

where $\sum\limits_{*}$ is sum over all odd $b_i > b$, such that $\frac{g}{b_i}$ is a positive integer.   □

6.3   Truncation and normal approximation

Truncation: From Section 6.2, $n = n'$ and $S(t) = S(n - t)$ except for $t = 0$, $n/2$. So for $\mathcal{P}_j \neq S(0)$, $S(n/2)$, we can define $\mathcal{A}_j$ such that

$$
\mathcal{P}_j = \{x : x \in \mathcal{A}_j \text{ or } n - x \in \mathcal{A}_j\} \text{ and } \#\mathcal{A}_j = \frac{1}{2}\#\mathcal{P}_j. \tag{6.11}
$$

For any sequence of random variables $b = \{b_l\}_{l \geq 0}$, define for $\mathcal{P}_j \in J_{2k}$,

$$
\beta_{b,k}(j) = \prod_{t \in \mathcal{A}_j} \Big|\frac{1}{\sqrt{n}}\sum_{l=0}^{n-1} b_l \omega^{tl}\Big|^2, \text{ where } \omega = \exp\Big(\frac{2\pi i}{n}\Big). \tag{6.12}
$$

Suppose $\{x_l\}_{l \geq 0}$ are independent, mean zero and variance one random variables. For each $n \geq 1$, define a triangular array of centered random variables $\{\bar{x}_l^{(n)}\}_{0 \leq l < n}$ by

$$
\bar{x}_l = \bar{x}_l^{(n)} = x_l \mathbb{I}_{|x_l| \leq n^{1/\gamma}} - E x_l \mathbb{I}_{|x_l| \leq n^{1/\gamma}}.
$$

Now, recall from Lemma 6.2.1, $\#J_{2g} = n_{2g} \approx \frac{n}{2g}$ for $n = k^g + 1$. Without loss of generality, assume that $\mathcal{P}_j \in J_{2g}$ for $1 \leq j \leq q = \frac{n}{2g}$. We then have the following lemma.

Lemma 6.3.1. Assume $E|x_l|^\gamma < \infty$ for some $\gamma > 2$. Then, almost surely,

$$
\max_{1 \leq j \leq q} (\beta_{x,g}(j))^{1/2g} - \max_{1 \leq j \leq q} (\beta_{\bar{x},g}(j))^{1/2g} = o(1).
$$

Proof. Since $\sum_{l=0}^{n-1} \omega^{tl} = 0$ for $0 < t < n$, it follows that $\beta_{\bar{x},n}(j) = \beta_{\tilde{x},n}(j)$ where

$$
\tilde{x}_l = \tilde{x}_l^{(n)} = \bar{x}_l + E x_l \mathbb{I}_{|x_l| \leq n^{1/\gamma}} = x_l \mathbb{I}_{|x_l| \leq n^{1/\gamma}}.
$$

By the Borel-Cantelli lemma, $\sum_{t=0}^{\infty} |x_t| \mathbb{I}_{|x_t| > t^{1/\gamma}}$ is finite a.s. and has only finitely many non-zero terms. Thus, there exists a random positive integer $N(\omega)$ such that

$$\sum_{t=0}^{n} |x_t - \tilde{x}_t| = \sum_{t=0}^{n} |x_t| \mathbb{I}_{|x_t|>n^{1/\gamma}} \le \sum_{t=0}^{\infty} |x_t| \mathbb{I}_{|x_t|>t^{1/\gamma}} = \sum_{t=0}^{N(\omega)} |x_t| \mathbb{I}_{|x_t|>t^{1/\gamma}}.$$

$$(6.13)$$

It follows that for $n \ge \{N, |x_1|^\gamma, \ldots, |x_N|^\gamma\}$, the left side of (6.13) is zero. Consequently, for all n sufficiently large,

$$\beta_{x,n}(j) = \beta_{\tilde{x},n}(j) = \beta_{\bar{x},n}(j) \text{ a.s. for all } j, \qquad (6.14)$$

and the assertion follows immediately. $\qquad\square$

Normal approximation: For $d \ge 1$, and any distinct integers $i_1, i_2, \ldots, i_d$, from $\{1, 2, \ldots, \lceil \frac{n-1}{2} \rceil\}$, define

$$v_{2d}(l) = \left(\cos \left(\frac{2\pi i_j l}{n} \right), \sin \left(\frac{2\pi i_j l}{n} \right) : 1 \le j \le d \right)^T, \quad l \in \mathbb{Z}_n.$$

Let $\phi_\Sigma(\cdot)$ denote the density of the $2d$-dimensional normal random vector which has mean zero and covariance matrix Σ. Let I_{2d} be the identity matrix of order $2d$. The following lemma is from Davis and Mikosch (1999). It is also a consequence of Lemma 5.2.2.

Lemma 6.3.2 (Davis and Mikosch (1999)). Let $\{x_t\}$ be i.i.d. random variables with $\mathrm{E}[x_0] = 0$, $\mathrm{E}[x_0]^2 = 1$, and $\mathrm{E}|x_0|^\gamma < \infty$ for some $\gamma > 2$. Let $\tilde{p}_n$ be the density function of

$$2^{1/2} \frac{1}{\sqrt{n}} \sum_{t=1}^{n} (\bar{x}_t + \sigma_n N_t) v_{2d}(t),$$

where $\{N_t\}$ is independent of $\{x_t\}$ and $\sigma_n^2 = \mathrm{Var}(\bar{x}_t) s_n^2$, for some sequence $\{s_n\}$. If $n^{-2c} \ln n < s_n^2 \le 1$ with $c = 1/2 - (1 - \delta)/\gamma$ for arbitrarily small $\delta > 0$, then uniformly for $|x|^3 = o_d(\min(n^c, n^{1/2-1/s}))$,

$$\tilde{p}_n(x) = \phi_{(1+\sigma_n^2)I_{2d}}(x)(1 + o(1)).$$

We shall use this lemma also in Section 8.2. The following corollary is similar to Lemma 5.2.3.

Corollary 6.3.3. Let $\gamma > 2$ and $\sigma_n^2 = n^{-c}$ where c is as in Lemma 6.3.2. Then for any measurable $B \subseteq \mathbb{R}^{2d}$,

$$\left| \int_B \tilde{p}_n(x) dx - \int_B \phi_{(1+\sigma_n^2)I_{2d}}(x) dx \right| \le \varepsilon_n \int_B \phi_{(1+\sigma_n^2)I_{2d}}(x) dx + O_d(\exp(-n^\eta)),$$

where $\varepsilon_n \to 0$ as $n \to \infty$ and $\eta > 0$. The above holds uniformly over all the d-tuples of distinct integers $1 \le i_1 < i_2 < \cdots < i_d \le \lceil \frac{n-1}{2} \rceil$.

Proof. Set $r = n^\alpha$ where $0 < \alpha < 1/2 - 1/\gamma$. Using Lemma 6.3.2, we have

$$\left| \int_B \tilde{p}_n(x)dx - \int_B \varphi_{(1+\sigma_n^2)I_{2d}}(x)dx \right|$$

$$\leq \left| \int_{B \cap \{\|x\| \leq r\}} \tilde{p}_n(x)dx - \int_{B \cap \{\|x\| \leq r\}} \varphi_{(1+\sigma_n^2)I_{2d}}(x)dx \right|$$

$$+ \int_{B \cap \{\|x\| > r\}} \tilde{p}_n(x)dx + \int_{B \cap \{\|x\| > r\}} \varphi_{(1+\sigma_n^2)I_{2d}}(x)dx$$

$$\leq \varepsilon_n \int_{B \cap \{\|x\| \leq r\}} \varphi_{(1+\sigma_n^2)I_{2d}}(x)dx + \int_{\{\|x\| > r\}} \tilde{p}_n(x)dx + \int_{\{\|x\| > r\}} \varphi_{(1+\sigma_n^2)I_{2d}}(x)dx$$

$$= T_1 + T_2 + T_3 \quad \text{(say)}.$$

Let $v_{2d}^{(j)}(l)$ denote the j-th coordinate of $v_{2d}(l)$, $1 \leq j \leq 2d$. Then, using the normal tail bound, $P\left(|N(0, \sigma^2)| > x\right) \leq 2e^{-x/\sigma}$ for $x > 0$,

$$T_2 = \int_{\{\|x\| > r\}} \tilde{p}_n(x)dx = P\left(\left\| 2^{1/2} \frac{1}{\sqrt{n}} \sum_{l=0}^{n-1} (\bar{a}_l + \sigma_n N_l) v_{2d}(l) \right\| > r \right)$$

$$\leq 2d \max_{1 \leq j \leq 2d} P\left(\left| 2^{1/2} \frac{1}{\sqrt{n}} \sum_{l=0}^{n-1} (\bar{a}_l + \sigma_n N_l) v_d^{(j)}(l) \right| > r/(2d) \right)$$

$$\leq 2d \max_{1 \leq j \leq 2d} P\left(\left| \frac{1}{\sqrt{n}} \sum_{l=0}^{n-1} \bar{a}_l v_d^{(j)}(l) \right| > r/(4\sqrt{2}d) \right) + 4d \exp\left(- rn^{c/2}/(4\sqrt{2}d) \right).$$

Note that $\bar{a}_l v_d^{(j)}(l), 0 \leq l < n$ are independent, have mean zero, variance at most one, and are bounded by $2n^{1/\gamma}$. Therefore, by applying Bernstein's inequality and simplifying, for some constant $K > 0$,

$$P\left(\left| \frac{1}{\sqrt{n}} \sum_{l=0}^{n-1} \bar{a}_l v_d^{(j)}(l) \right| > r/4\sqrt{2}d \right) \leq \exp(-Kr^2).$$

Further,

$$T_3 = \int_{\{\|x\| > r\}} \varphi_{(1+\sigma_n^2)I_{2d}}(x)dx \leq 4d \exp(-r/4d).$$

Combining the above estimates finishes the proof. $\qquad \square$

6.4   Spectral radius of the k-circulant

Now we finally arrive at the main result of this chapter.

Theorem 6.4.1 (Bose et al. (2012a)). Suppose $\{x_i\}_{i \geq 0}$ is an i.i.d. sequence of random variables with mean zero, variance 1, and $E\,|x_i|^\gamma < \infty$ for some $\gamma > 2$. If $n = k^g + 1$ for some fixed positive integer g, then

$$\frac{\mathrm{sp}(\frac{1}{\sqrt{n}} A_{k,n}) - d_q}{c_q} \xrightarrow{D} \Lambda, \quad \text{as } n \to \infty,$$

where $q = q_n = \frac{n}{2g}$,

$$c_n = \frac{1}{2g^{1/2}(\ln n)^{1/2}},$$

$$d_n = \frac{\ln C_g - \frac{g-1}{2}\ln g}{2g^{1/2}(\ln n)^{1/2}} + \left(\frac{\ln n}{g}\right)^{1/2}\left[1 + \frac{(g-1)\ln\ln n}{4\ln n}\right], \text{ and}$$

$$C_g = \frac{1}{\sqrt{g}}(2\pi)^{\frac{g-1}{2}}.$$

To establish the theorem we shall use the following lemmata whose proofs are given later.

Recall that $\{\beta_{x,g}(t)^{1/2g}\}$ are the eigenvalues corresponding to the set of partitions which have cardinality $2g$. We derive the behavior of the maximum of these eigenvalues in Lemma 6.4.2. Then using the results of Lemma 6.4.3, we show that the maximum of the remaining eigenvalues is negligible compared to the above.

Lemma 6.4.2.

$$\frac{\max_{1 \leq t \leq q} \beta_{x,g}(t)^{1/2g} - d_q}{c_q} \xrightarrow{D} \Lambda, \tag{6.15}$$

where d_q, c_q are as in Corollary 6.1.3, $q = q_n = \frac{n}{2g} - k_n$ and $\frac{k_n}{n} \to 0$ as $n \to \infty$. As a consequence,

$$\frac{\max_{1 \leq t \leq q} \beta_{x,g}(t)^{1/2g} - d_{n/2g}}{c_{n/2g}} \xrightarrow{D} \Lambda. \tag{6.16}$$

The next lemma is technical. Let

$$c_n(l) = \frac{1}{2l^{1/2}(\ln n)^{1/2}},$$

$$d_n(l) = \frac{\ln C_l - \frac{l-1}{2}\ln l}{2l^{1/2}(\ln n)^{1/2}} + \left(\frac{\ln n}{l}\right)^{1/2}\left[1 + \frac{(l-1)\ln\ln n}{4\ln n}\right],$$

$$C_l = \frac{1}{\sqrt{l}}(2\pi)^{\frac{l-1}{2}}, \text{ and}$$

$$c_{n_{2j}} = c_{n_{2j}}(j), \ d_{n_{2j}} = d_{n_{2j}}(j), \ c_{n/2g} = c_{n/2g}(g) \text{ and } d_{n/2g} = d_{n/2g}(g).$$

Lemma 6.4.3. Let $n = k^g + 1$. If $j < g$ and for some $a > 0$, $2jn_{2j} = (k^j + 1)(1 + n^{-a}) \approx n^{\frac{j}{g}}$ or is bounded, then there exists a constant $K = K(j, g) \geq 0$ such that

$$\frac{c_{n/2g}}{c_{n_{2j}}} \to K \quad \text{and} \quad \frac{d_{n/2g} - d_{n_{2j}}}{c_{n_{2j}}} \to \infty \text{ as } n \to \infty.$$

Proof of Theorem 6.4.1. If $\#\mathcal{P}_i = j$, then the eigenvalues corresponding to $\mathcal{P}_i$'s are the j-th roots of y_i and hence these eigenvalues have the same modulus. From Lemma 6.2.1, the possible values of $\#\mathcal{P}_i$ are $\{1, 2, 2g \text{ and } 2g/b \ (3 \leq b < g, b \text{ odd}, \frac{g}{b} \in \mathbb{Z})\}$. Recall from (6.12) that $\beta_{x,j}(i)$ is the modulus of the eigenvalue associated with the partition set $\mathcal{P}_i$, where $\#\mathcal{P}_i = 2j$.

Normally distributed case: In case of normally distributed entries, it easily follows that $\beta_{x,j}(i)$ is the product of j exponential random variables, and they are independent as i takes n_{2j}-many distinct values. So from Corollary 6.1.3, if $n_{2j} \to \infty$ then the maximum of $\beta_{x,j}(k)^{1/2j}$ has a Gumbel limit.

Non-normal case: For general entries the proof of Lemma 6.4.2 also implies that

$$\max_{1 \leq k \leq n_{2j}} \frac{\beta_{x,j}(k)^{1/2j} - d_{n_{2j}}}{c_{n_{2j}}} \xrightarrow{\mathcal{D}} \Lambda \text{ as } n_{2j} \to \infty, \tag{6.17}$$

where $c_{n_{2j}}$ and $d_{n_{2j}}$ are as above.

Now let

$$x_n = c_n x + d_n, \quad q = q(n) = \frac{n}{2g} \quad \text{and} \quad \mathcal{B} = \{b : b \text{ odd}, 3 \leq b < g, \frac{g}{b} \in \mathbb{Z}\}.$$

Then

$$P\left(\text{sp}(\frac{1}{\sqrt{n}}A_{k,n}) > x_q\right) \geq P\left(\max_{j:\mathcal{P}_j \in J_{2g}} \beta_{x,g}(j)^{1/2g} > x_q\right),$$

and

$$P\left(\text{sp}(\frac{1}{\sqrt{n}}A_{k,n}) > x_q\right)$$

$$\leq P\left(\max_{j:\mathcal{P}_j \in J_{2g}} \beta_{x,g}(j)^{1/2g} > x_q\right) + \sum_{b \in \mathcal{B}} P\left(\max_{j:\mathcal{P}_j \in J_{\frac{2g}{b}}} \beta_{x,\frac{g}{b}}(j)^{b/2g} > x_q\right)$$

$$+ P\left(|\frac{1}{\sqrt{n}}\sum_{l=0}^{n-1} x_l| > x_q\right) + P\left(|\frac{1}{\sqrt{n}}\sum_{l=0}^{n-1}(-1)^l x_l| > x_q\right)$$

$$+ P\left(\max_{j:\mathcal{P}_j \in J_2} \beta_{x,2}(j)^{1/2} > x_q\right)$$

$$=: \quad A + B + C + D + E.$$

We first verify that B, C, D, E are negligible. From Lemma 6.2.1, the term D appears only when $\frac{n}{2} \in \mathbb{Z}$, and the term E appears only if g is even and in that case J_2 contains only one element. It is easy to see that C, D and E tend to zero since we are taking the maximum of a single element.

Note that B is a sum of finitely many terms. Now suppose for $b \in \mathcal{B}$, we have some finite K_b such that

$$\frac{c_{n/2g}}{c_{n_{2g/b}}} \to K_b \quad \text{and} \quad \frac{d_{n/2g} - d_{n_{2g/b}}}{c_{n_{2g/b}}} \to \infty \text{ as } n \to \infty. \tag{6.18}$$

Observations (6.17) and (6.18) imply B goes to zero. So it remains to verify (6.18) for $b \in \mathcal{B}$. But this follows from Lemma 6.2.1(e) and Lemma 6.4.3.

Now the limit in A can be identified using Lemma 6.4.2 and that finishes the proof of the theorem. $\qquad\square$

Proof of Lemma 6.4.2. First assume that $\{x_l\}_{l\geq0}$ are i.i.d. standard normal. Let $\{E_j\}_{j\geq1}$ be i.i.d. standard exponentials. By Lemma 3.3.2, it easily follows that

$$P\left(\max_{1\leq t\leq q}(\beta_{x,g}(t))^{1/2g} > c_q x + d_q\right)$$
$$= P\left(\left(E_{g(j-1)+1}E_{g(j-1)+2}\cdots E_{gj}\right)^{1/2g} > c_q x + d_q \text{ for some } 1 \leq j \leq q\right).$$

The lemma then follows in this *special case* from Corollary 6.1.3.

For the general case, we break the proof into the following three steps and make use of the two results from Section 6.3. The proof of these three steps will be given later. Fix $x \in \mathbb{R}$.

Step 1: Claim:

$$\lim_{n\to\infty}[Q_1^{(n)} - Q_2^{(n)}] = 0, \tag{6.19}$$

where

$$Q_1^{(n)} := P\left(\max_{1\leq j\leq q}(\beta_{\bar{x}+\sigma_n N,g}(j))^{1/2g} > c_q x + d_q\right),$$
$$Q_2^{(n)} := P\left(\max_{1\leq j\leq q}(1+\sigma_n^2)^{1/2}\left(E_{g(j-1)+1}E_{g(j-1)+2}\cdots E_{gj}\right)^{1/2g} > c_q x + d_q\right),$$

and $\{N_l\}_{l\geq0}$ is a sequence of i.i.d. standard normal random variables.

Step 2: Claim:

$$\frac{\max_{1\leq j\leq q}(\beta_{\bar{x}+\sigma_n N,g}(j))^{1/2g} - d_q}{c_q} \overset{\mathcal{D}}{\to} \Lambda. \tag{6.20}$$

Step 3: Claim:

$$\frac{\max_{1\leq t\leq q}\beta_{\bar{x},g}(t)^{1/2g} - d_q}{c_q} \overset{\mathcal{D}}{\to} \Lambda. \tag{6.21}$$

Now combining Lemma 6.3.1 and (6.21) we can conclude that

$$\frac{\max_{1\leq t\leq q}\beta_{x,g}(t)^{1/2g} - d_q}{c_q} \overset{\mathcal{D}}{\to} \Lambda.$$

This completes the proof of the first part, namely of (6.15) of the lemma. By convergence of types result, the second part, namely, (6.16) follows since the following hold. We omit the tedious algebraic details.

$$\frac{c_q}{c_{n/2g}} \to 1 \text{ and } \frac{d_q - d_{n/2g}}{c_q} \to 0, \text{ as } n \to \infty. \tag{6.22}$$

Proof of Step 1: We approximate $Q_1^{(n)}$ by the simpler quantity $Q_2^{(n)}$ using Bonferroni inequality. For all $m \geq 1$,

$$\sum_{j=1}^{2m} (-1)^{j-1} S_{j,n} \leq Q_1^{(n)} \leq \sum_{j=1}^{2m-1} (-1)^{j-1} S_{j,n}, \tag{6.23}$$

where

$$S_{j,n} = \sum_{1 \leq t_1 < t_2 < \ldots < t_j \leq q} P\left((\beta_{\bar{x} + \sigma_n N, g}(t_i))^{1/2g} > c_q x + d_q, \ i = 1, \ldots, j \right).$$

Similarly, we have

$$\sum_{j=1}^{2m} (-1)^{j-1} T_{j,n} \leq Q_2^{(n)} \leq \sum_{j=1}^{2m-1} (-1)^{j-1} T_{j,n}, \tag{6.24}$$

where

$$T_{j,n} = \sum_{\mathcal{T}} P\left(\sqrt{1 + \sigma_n^2} (E_{g(t_i-1)+1} E_{g(t_i-1)+2} \cdots E_{gt_i})^{\frac{1}{2g}} > c_q x + d_q, 1 \leq i \leq j \right)$$

and $\mathcal{T} = \{1 \leq t_1 < t_2 < \cdots < t_j \leq q\}$. Therefore, the difference between $Q_1^{(n)}$ and $Q_2^{(n)}$ can be bounded as follows:

$$\sum_{j=1}^{2m} (-1)^{j-1} (S_{j,n} - T_{j,n}) - T_{2m+1,n} \quad \leq \quad Q_1^{(n)} - Q_2^{(n)} \tag{6.25}$$

$$\leq \quad \sum_{j=1}^{2m-1} (-1)^{j-1} (S_{j,n} - T_{j,n}) + T_{2m,n},$$

for each $m \geq 1$. By independence and Lemma 6.1.4, there exists $K = K(x)$ such that

$$T_{j,n} \leq \binom{n}{j} \frac{K^j}{n^j} \leq \frac{K^j}{j!} \quad \text{for all } n, j \geq 1. \tag{6.26}$$

Consequently, $\lim_{j \to \infty} \limsup_n T_{j,n} = 0$.

Now fix $j \geq 1$. Let us bound the difference between $S_{j,n}$ and $T_{j,n}$. Let

$\mathcal{A}_t$ defined in (6.11) be represented as $\mathcal{A}_t = \{e_t^1, e_t^2, \ldots, e_t^g\}$. Also note that $e_t^1, e_t^2, \ldots, e_t^g \in \{1, 2, \ldots, \lfloor \frac{n}{2} \rfloor\}$. For $1 \le t_1 < t_2 < \cdots < t_j \le q$, define

$$v_{2gj}(l)$$
$$= \left(\cos\left(\frac{2\pi l e_{t_k}^1}{n}\right), \sin\left(\frac{2\pi l e_{t_k}^1}{n}\right), \cos\left(\frac{2\pi l e_{t_k}^2}{n}\right), \ldots, \sin\left(\frac{2\pi l e_{t_k}^g}{n}\right); \ 1 \le k \le j \right).$$

Note that $\{e_{t_k}^1, \ldots, e_{t_k}^g : 1 \le k \le j\}$ is a set of distinct integers in $\{1, 2, \ldots, \lfloor \frac{n}{2} \rfloor\}$. Then

$$P\left((\beta_{\bar{x}+\sigma_n N, g}(t_i))^{1/2g} > c_q x + d_q, i = 1, \ldots, j\right)$$
$$= P\left(2^{1/2} \frac{1}{\sqrt{n}} \sum_{l=0}^{n-1} (\bar{x}_l + \sigma_n N_l) v_{2gj}(l) \in B_n^{(j)}\right),$$

where

$$B_n^{(j)} := \left\{y \in \mathbb{R}^{2gj} : \prod_{l=1}^{g} (y_{2gt+2l-1}^2 + y_{2gt+2l}^2)^{1/2g} > 2^{1/2}(c_q x + d_q); 0 \le t < j\right\}.$$

By Corollary 6.3.3 and the fact that $N_1^2 + N_2^2 \overset{D}{=} 2E_1$, we deduce that uniformly over all the d-tuples $1 \le t_1 < t_2 < \cdots < t_j \le q$,

$$\left| P\left(2^{1/2} \frac{1}{\sqrt{n}} \sum_{l=0}^{n-1} (\bar{x}_l + \sigma_n N_l) v_{2gj}(l) \in B_n^{(j)}\right) \right.$$
$$\left. - P\left((1 + \sigma_n^2)^{1/2} \left(\prod_{i=1}^{g} E_{g(t_m-1)+i}\right)^{1/2g} > c_q x + d_q, 1 \le m \le j\right) \right|$$
$$\le \epsilon_n P\left((1 + \sigma_n^2)^{\frac{1}{2}} \left(E_{g(t_m-1)+1} E_{g(t_m-1)+2} \cdots E_{g t_m}\right)^{\frac{1}{2g}} > c_q x + d_q, 1 \le m \le j\right)$$
$$+ O(\exp(-n^\eta)).$$

Therefore, as $n \to \infty$,

$$|S_{j,n} - T_{j,n}| \le \epsilon_n T_{j,n} + \binom{n}{j} O(\exp(-n^\eta)) \le \epsilon_n \frac{K^j}{j!} + o(1) \to 0, \qquad (6.27)$$

where $O(\cdot)$ and $o(\cdot)$ are uniform over j. Hence, using (6.23), (6.24), (6.26) and (6.27), we have

$$\limsup_{n} |Q_1^{(n)} - Q_2^{(n)}| \le \limsup_{n} T_{2m+1,n} + \limsup_{n} T_{2m,n}, \quad \text{for each } m \ge 1.$$

Letting $m \to \infty$, we conclude that $\lim_{n\to\infty}[Q_1^{(n)} - Q_2^{(n)}] = 0$. This completes the proof of Step 1.

Proof of Step 2: Since by Corollary 6.1.3,

$$\max_{1 \le j \le q} \left(E_{g(j-1)+1} E_{g(j-1)+2} \cdots E_{gj}\right)^{1/2g} = O_p((\ln n)^{1/2}) \quad \text{and} \quad \sigma_n^2 = n^{-c},$$

it follows that

$$\frac{(1 + \sigma_n^2)^{1/2} \max_{1 \le j \le q} \left(E_{g(j-1)+1} E_{g(j-1)+2} \cdots E_{gj} \right)^{1/2g} - d_q}{c_q} \xrightarrow{D} \Lambda,$$

and consequently

$$\frac{\max_{1 \le j \le q} (\beta_{\bar{x} + \sigma_n N, g}(j))^{1/2g} - d_q}{c_q} \xrightarrow{D} \Lambda.$$

This completes the proof of Step 2.

Proof of Step 3: In view of (6.20), it suffices to show that

$$\max_{1 \le j \le q} (\beta_{\bar{x} + \sigma_n N, g}(j))^{1/2g} - \max_{1 \le j \le q} (\beta_{\bar{x}, g}(j))^{1/2g} = o_p(c_q).$$

Note that

$$\beta_{\bar{x} + \sigma_n N, g}(j) = \prod_{k=1}^{g} \left| \frac{1}{\sqrt{n}} \sum_{l=0}^{n-1} (\bar{x}_l + \sigma_n N_l) \omega^{le_j^k} \right|^2 = \prod_{k=1}^{g} |\alpha_{j,k}|^2 \quad \text{(say)},$$

and

$$\beta_{\bar{x}, g}(j) = \prod_{k=1}^{g} \left| \frac{1}{\sqrt{n}} \sum_{l=0}^{n-1} \bar{x}_l \omega^{le_j^k} \right|^2 = \prod_{k=1}^{g} |\gamma_{j,k}|^2 \quad \text{(say)}.$$

Now by the inequality,

$$\left| \prod_{i=1}^{g} a_i - \prod_{i=1}^{g} b_i \right| \le \sum_{j=1}^{g} \left(\prod_{i=1}^{j-1} b_i \right) |a_j - b_j| \left(\prod_{i=j+1}^{g} a_i \right) \tag{6.28}$$

for non-negative numbers $\{a_i\}$ and $\{b_i\}$, we have

$$\left| \beta_{\bar{x} + \sigma_n N, g}(j) - \beta_{\bar{x}, g}(j) \right|$$

$$\le \sum_{k=1}^{g} |\gamma_{j,1}|^2 \cdots |\gamma_{j,k-1}|^2 \Big| |\alpha_{j,k}|^2 - |\gamma_{j,k}|^2 \Big| |\alpha_{j,k+1}|^2 \cdots |\alpha_{j,g}|^2.$$

For any sequence of random variables $\{X_n\}_{n \ge 0}$, define

$$M_n(X) := \max_{1 \le t \le n} \left| \frac{1}{\sqrt{n}} \sum_{l=0}^{n-1} X_l \omega^{tl} \right|.$$

As a trivial consequence of Theorem 2.1 of Davis and Mikosch (1999) (see Theorem 11.3.2 in Appendix), we have

$$M_n^2(\sigma_n N) = O_p(\sigma_n \ln n) \quad \text{and} \quad M_n^2(\bar{x} + \sigma_n N) = O_p(\ln n).$$

Therefore $\left|\alpha_{j,k}\right| = O_p(\sqrt{\ln n})$. Now

$$\left|\gamma_{j,k}\right| \leq \left|\alpha_{j,k}\right| + \sigma_n \left|\frac{1}{\sqrt{n}}\sum_{l=0}^{n-1} N_l \omega^{le_j^k}\right|,$$

and therefore $\left|\gamma_{j,k}\right| = (1 + \sigma_n)O_p(\sqrt{\ln n}) = O_p(\sqrt{\ln n})$. So we have

$$\left|\max_{1 \leq j \leq q} \beta_{\bar{x}+\sigma_n N, g}(j) - \max_{1 \leq j \leq q} \beta_{\bar{x}, g}(j)\right|$$

$$\leq \max_{1 \leq j \leq q} \left|\beta_{\bar{x}+\sigma_n N, g}(j) - \beta_{\bar{x}, g}(j)\right|$$

$$\leq \max_{1 \leq j \leq q} \sum_{k=1}^{g} \left(O_p(\ln n)\right)^{g-1} \left|\alpha_{j,k} - \gamma_{j,k}\right| \left(\left|\alpha_{j,k}\right| + \left|\gamma_{j,k}\right|\right)$$

$$\leq O_p(\ln n)^{g-1} O_p(\sqrt{\ln n}) \max_{1 \leq j \leq q} \sum_{k=1}^{g} \left|\alpha_{j,k} - \gamma_{j,k}\right|$$

$$\leq O_p(\ln n)^{g-\frac{1}{2}} g\sigma_n M_n(N)$$

$$\leq o_p\left(n^{-c/4}(\ln n)^g\right).$$

Hence

$$\left|\max_{1 \leq j \leq q} (\beta_{\bar{x}+\sigma_n N, g}(j))^{1/2g} - \max_{1 \leq j \leq q} (\beta_{\bar{x}, g}(j))^{1/2g}\right|$$

$$\leq \left|\max_{1 \leq j \leq q} \beta_{\bar{x}+\sigma_n N, g}(j) - \max_{1 \leq j \leq q} \beta_{\bar{x}, g}(j)\right| \frac{1}{2g\xi^{1-1/2g}},$$

where ξ lies between $\max_{1 \leq j \leq q} \beta_{\bar{x}+\sigma_n N, g}(j)$ and $\max_{1 \leq j \leq q} \beta_{\bar{x}, g}(j)$. We know that

$$\frac{\max_{1 \leq j \leq q} \beta_{\bar{x}+\sigma_n N, g}(j)}{(\ln n)^g} \xrightarrow{\mathcal{P}} 1 \text{ and}$$

$$\frac{\left|\max_{1 \leq j \leq q} \beta_{\bar{x}+\sigma_n N, g}(j) - \max_{1 \leq j \leq q} \beta_{\bar{x}, g}(j)\right|}{(\ln n)^g} \xrightarrow{\mathcal{P}} 0.$$

Therefore

$$\frac{\max_{1 \leq j \leq q} \beta_{\bar{x}, g}}{(\ln n)^g}$$

$$= \frac{\max_{1 \leq j \leq q} \beta_{\bar{x}+\sigma_n N, g}(j)}{(\ln n)^g} + \frac{\max_{1 \leq j \leq q} \beta_{\bar{x}, g}(j) - \max_{1 \leq j \leq q} \beta_{\bar{x}+\sigma_n N, g}(j)}{(\ln n)^g} \xrightarrow{\mathcal{P}} 1.$$

Hence

$$\frac{\xi}{(\ln n)^g} \xrightarrow{\mathcal{P}} 1 \Rightarrow \frac{\xi^{1-1/2g}}{(\ln n)^{g(1-1/2g)}} \xrightarrow{\mathcal{P}} 1 \Rightarrow \frac{1}{\xi^{1-1/2g}} = O_p\left((\ln n)^{\frac{1}{2}-g}\right).$$

Combining all these we have

$$\left| \max_{1 \le j \le q} \beta_{\bar{x}+\sigma_n N, g}(j)^{1/2g} - \max_{1 \le j \le q} \beta_{\bar{x}, g}(j)^{1/2g} \right| \le o_p\big(n^{-c/4}(\ln n)^g\big)O_p\big((\ln n)^{\frac{1}{2}-g}\big)$$

$$= o_p(c_q).$$

This completes the proof of Step 3, and hence completes the proof of Lemma 6.4.2. □

Proof of Lemma 6.4.3. First observe that if n_j is finite then the result holds trivially. If $n_{2j} = \frac{(k^j+1)(1+n^{-a})}{2j}$ then

$$\ln n_{2j} = j \ln k + \Big(\frac{1}{n^a} + \frac{1}{n^{j/g}}\Big)(1+o(1)) - \ln 2j,$$

for some $a > 0$, and since $k = (n-1)^{\frac{1}{g}}$, we have

$$\frac{c_{n/2g}}{c_{n_{2j}}} \to \frac{j}{g} \quad \text{as } n \to \infty.$$

Similarly we get for some $a_0 > 0$,

$$\ln \ln n_{2j} = \ln \ln n^{\frac{j}{g}} + \Big(\frac{1}{n^{a_0} \ln n}\Big)(1+o(1)) - \ln 2j.$$

Now observe that $\frac{d_{n/2g} - d_{n_{2j}}}{c_{n_{2j}}}$ can be broken into the following three parts:

$$J_1 = 2j^{1/2}(\ln n_{2j})^{1/2}\Big[\frac{\ln C_g - \frac{g-1}{2}\ln g}{2g^{1/2}(\ln \frac{n}{2g})^{1/2}} - \frac{\ln C_j - \frac{j-1}{2}\ln j}{2j^{1/2}(\ln n_{2j})^{1/2}}\Big] \to m_1 \quad \text{(finite).}$$

$$J_2 = 2j^{1/2}(\ln n_{2j})^{1/2}\Big[\big(\frac{\ln n/2g}{g}\big)^{1/2} - \big(\frac{\ln n_{2j}}{j}\big)^{1/2}\Big] \to m_2 \quad \text{(finite).}$$

$$J_3 = 2j^{1/2}(\ln n_{2j})^{1/2}\Big[\frac{(g-1)\ln \ln n/2g}{4(g\ln n/2g)^{1/2}} - \frac{(j-1)\ln \ln n_{2j}}{4(j\ln n_{2j})^{1/2}}\Big]$$

$$= 2j^{1/2}(\ln n_{2j})^{1/2}\Big[\frac{(g-1)\ln \ln n/2g}{4(g\ln n/2g)^{1/2}} - \frac{(j-1)\sqrt{g}\ln \ln n_{2j}}{4j(\ln n/2g)^{1/2}} + o(1)\Big]$$

$$= \frac{j^{1/2}(\ln n_{2j})^{1/2}}{2(g\ln n/2g)^{1/2}}\Big[(g-1)\ln \ln n/2g - \frac{(j-1)g}{j}\ln \ln n_{2j} + o(1)\Big]$$

$$= \frac{j^{1/2}(\ln n_{2j})^{1/2}}{2(g\ln n/2g)^{1/2}}\Big[\big((g-1) - \frac{g(j-1)}{j}\big)\ln \ln n/2g + o(1)\Big] \to \infty \quad \text{(as } g > j\text{).}$$

Hence Lemma 6.4.3 is proven. □

6.4.1 *k*-circulant for $sn = k^g + 1$

We have seen in Chapter 4 that the LSD of $\frac{1}{\sqrt{n}}A_{k,n}$ exists when $k^g = sn - 1$ and $s = o(n^{p_1-1})$, where p_1 is the smallest prime factor of g. To derive a limit for the spectral radius, we need to strengthen this assumption slightly.

Theorem 6.4.4 (Bose et al. (2012a)). Suppose $\{x_l\}_{l \geq 0}$ is an i.i.d. sequence of random variables with mean zero, variance 1, and $E|x_l|^\gamma < \infty$ for some $\gamma > 2$. If $s \geq 1$, $sn = k^g + 1$ where $s = o(n^{p_1-1-\varepsilon})$, $0 < \varepsilon < p_1$, and p_1 is the smallest prime factor of g, then as $n \to \infty$,

$$\frac{\mathrm{sp}(\frac{1}{\sqrt{n}}A_{k,n}) - d_q}{c_q} \xrightarrow{\mathcal{D}} \Lambda,$$

where $q = q(n) = \frac{n}{2g}$. The constants c_n and d_n can be taken as follows:

$$d_n = \frac{\ln C_g - \frac{g-1}{2}\ln g}{2g^{1/2}(\ln n)^{1/2}} + \left(\frac{\ln n}{2g}\right)^{1/2}\left[1 + \frac{(g-1)\ln\ln n}{4\ln n}\right],$$

$$c_n = \frac{1}{2g^{1/2}(\ln n)^{1/2}}, \quad \text{and}$$

$$C_g = \frac{1}{\sqrt{g}}(2\pi)^{\frac{g-1}{2}}.$$

Note that the case $s = 1$ reduces to the previous theorem. The proof for the general case is along the same line except for certain modifications.

The condition $s = o(n^{p_1-1})$ implies $v_{k,n}/n \to 0$. But this is not enough to neglect certain terms. We need the stronger result $\frac{v_{k,n}}{n} = o(n^{-a_1})$ for some $a_1 > 0$ so that these terms are negligible in the log scale that we have. We omit the detailed proof of this but provide a brief heuristic explanation.

Since $s > 1$ it can be checked that (see proof of Lemma 1.4.4),

$$\#X\left(\frac{2g}{b}\right) \leq \gcd\left(k^{2g/b} - 1, \frac{k^g + 1}{s}\right)$$

$$\leq \gcd(k^{2g/b} - 1, k^g + 1). \tag{6.29}$$

Now recall $Z_{n,b}$ from (6.9) and observe that

$$\#\{x : x \in \mathbb{Z}_n \text{ and } xk^{2g/b} = x \bmod \left(\frac{k^g + 1}{s}\right)\} \geq \#Z_{n,b}. \tag{6.30}$$

From observations (6.29) and (6.30) it easily follows that Lemma 6.2.1(d) remains valid in this case also, that is,

$$X\left(\frac{2g}{b}\right) = \emptyset \quad \text{for } 2 \leq b < g, \quad b \text{ even}.$$

Moreover, if g is even then $X(\frac{2g}{g}) = X(2)$ is either empty or contains exactly two elements.

Further, it can be shown that (as in Lemma 6.2.1(e)), for b odd, $3 \le b \le g$,

$$1 \ge \frac{\#X(\frac{2g}{b})}{k^{g/b}+1} \ge 1 - n^{-\alpha}(1+o(1)),$$

for some $\alpha > 0$. Now the rest of the proof is as Theorem 6.4.1.

6.5   Exercises

1. Prove Corollary 6.1.3.

2. Prove Corollary 6.1.4.

3. Complete the proof of Theorem 6.4.4.

7

Maximum of scaled eigenvalues: dependent input

In this chapter we try to generalize the results of Chapters 5 and 6 on spectral radius to the situation where the input sequence is dependent. We take $\{x_n\}$ to be an infinite order moving average process, $x_n = \sum_{i=-\infty}^{\infty} a_i \varepsilon_{n-i}$, where $\{a_n; n \in \mathbb{Z}\}$ are non-random with $\sum_n |a_n| < \infty$, and $\{\varepsilon_i; i \in \mathbb{Z}\}$ are i.i.d. In this case the eigenvalues have unequal variances. So, we resort to scaling each eigenvalue by an appropriate quantity and then consider the distributional convergence of the maximum of these scaled eigenvalues of different circulant matrices. This scaling has the effect of (approximately) equalizing the variance of the eigenvalues. Similar scaling has been used in the study of the periodogram (see Walker (1965), Davis and Mikosch (1999), and Lin and Liu (2009)).

7.1 Dependent input with light tail

Let $\{x_n; n \geq 0\}$ be a two-sided moving average process,

$$x_n = \sum_{i=-\infty}^{\infty} a_i \varepsilon_{n-i}, \qquad (7.1)$$

where $\{a_n; n \in \mathbb{Z}\}$ are non-random, $\sum_n |a_n| < \infty$, and $\{\varepsilon_i; i \in \mathbb{Z}\}$ are i.i.d. random variables. Let $f(\omega)$, $\omega \in [0, 2\pi]$ be the spectral density of $\{x_n\}$. Note that

$$f \equiv \frac{\sigma^2}{2\pi} \text{ if } \{x_n\} \text{ is i.i.d. with mean } 0 \text{ and variance } \sigma^2.$$

We make the following assumption.

Assumption 7.1.1. $\{\varepsilon_i, \ i \in \mathbb{Z}\}$ are i.i.d. with $E(\varepsilon_i) = 0$, $E(\varepsilon_i^2) = 1$, $E|\varepsilon_i|^{2+\delta} < \infty$ for some $\delta > 0$,

$$\sum_{j=-\infty}^{\infty} |a_j| |j|^{1/2} < \infty, \text{ and } f(\omega) > 0 \text{ for all } \omega \in [0, 2\pi].$$

7.2 Reverse circulant and circulant

Define $M(\cdot, f)$ for the reverse circulant matrix as follows:

$$M(\tfrac{1}{\sqrt{n}}RC_n, f) = \max_{1 \le k < \frac{n}{2}} \frac{|\lambda_k|}{\sqrt{2\pi f(\omega_k)}},$$

where f is the spectral density corresponding to $\{x_n\}$, λ_k are the eigenvalues of $\frac{1}{\sqrt{n}}RC_n$ as defined in Section 1.3, and $\omega_k = \frac{2\pi k}{n}$. Note that $M(\frac{1}{\sqrt{n}}C_n, f)$ is defined similarly and satisfies $M(\frac{1}{\sqrt{n}}RC_n, f) = M(\frac{1}{\sqrt{n}}C_n, f)$. Note that λ_0 is not included in the definition of $M(\cdot, f)$. When $E(\varepsilon_0) = \mu = 0$, λ_0 does not affect the limiting behavior of $M(\frac{1}{\sqrt{n}}RC_n, f)$. However if $\mu \neq 0$, then it does. See Remark 7.2.1. In the next theorem we assume $\mu = 0$.

Theorem 7.2.1 (Bose et al. (2011c)). Let $\{x_n\}$ be the two-sided moving average process defined in (7.1) and which satisfies Assumption 7.1.1. Then

$$\frac{M(\frac{1}{\sqrt{n}}RC_n, f) - d_q}{c_q} \xrightarrow{D} \Lambda,$$

where $q = q(n) = \lfloor \frac{n-1}{2} \rfloor$, $d_q = \sqrt{\ln q}$ and $c_q = \frac{1}{2\sqrt{\ln q}}$. The same result continues to hold for $M(\frac{1}{\sqrt{n}}C_n, f)$.

We need the following lemma. This lemma is an approximation result which is a stronger version of Theorem 3 of Walker (1965). We will first use this result in the proof of Theorem 7.2.1 but not in full force. We will again use it in Section 7.4.

Lemma 7.2.2. Let $\{x_n\}$ be a two-sided moving average process as defined in (7.1) and which satisfies Assumption 7.1.1. Then

$$\max_{1 \le k < \frac{n}{2}} \left| \frac{I_{x,n}(\omega_k)}{2\pi f(\omega_k)} - I_{\varepsilon,n}(\omega_k) \right| = o_p(n^{-1/4}\sqrt{\ln n}),$$

where

$$I_{x,n}(\omega_k) = \frac{1}{n}|\sum_{t=0}^{n-1} x_t e^{it\omega_k}|^2, \quad I_{\varepsilon,n}(\omega_k) = \frac{1}{n}|\sum_{t=0}^{n-1} \varepsilon_t e^{it\omega_k}|^2, \text{ and } \omega_k = \frac{2\pi k}{n}.$$

Proof. First observe that $\min_{\omega \in [0, 2\pi]} f(\omega) > \alpha > 0$. Now for any r,

$$|\sum_{t=1}^{r} \varepsilon_t e^{i\omega t}|^2 = \sum_{s=-r}^{r} e^{i\omega s} \sum_{t=1}^{r-|s|} \varepsilon_t \varepsilon_{t+|s|}$$

$$\le \sum_{s=-r}^{r} |\sum_{t=1}^{r-|s|} \varepsilon_t \varepsilon_{t+|s|}|.$$

Hence

$$
\begin{aligned}
\mathrm{E}\Big[\max_{0\le\omega\le\pi}\big|\sum_{t=1}^{r}\varepsilon_t e^{i\omega t}\big|^2\Big] &\le \mathrm{E}\Big(\sum_{t=1}^{r}\varepsilon_t^2\Big)+2\sum_{s=1}^{r-1}\Big[\mathrm{E}\big(\sum_{t=1}^{r-s}\varepsilon_t\varepsilon_{t+s}\big)^2\Big]^{1/2} \\
&= r+2\sum_{s=1}^{r-1}(r-s)^{1/2} \\
&\le r+2\int_{1}^{r}x^{1/2}dx \\
&\le Kr^{3/2},
\end{aligned}
\tag{7.2}
$$

where K is a constant independent of r. So

$$
\mathrm{E}\Big[\max_{0\le\omega\le\pi}\big|\sum_{t=1}^{r}\varepsilon_t e^{i\omega t}\big|\Big]\le K^{1/2}r^{3/4}.
\tag{7.3}
$$

Note that (7.3) continues to hold if the limits of summation for t are replaced by $1+p$ and $r+p$, where p is an arbitrary (positive or negative) integer. Let

$$
\begin{aligned}
J_{x,n} &= \frac{1}{\sqrt{n}}\sum_{t=0}^{n-1}x_t e^{i\omega t}, \quad J_{\varepsilon,n}=\frac{1}{\sqrt{n}}\sum_{t=0}^{n-1}\varepsilon_t e^{i\omega t}, \\
R_n(\omega) &= J_{x,n}(\omega)-A(\omega)J_{\varepsilon,n}(\omega), \\
T_n(\omega) &= I_{x,n}(\omega)-|A(\omega)|^2 I_{\varepsilon,n}(\omega), \quad \text{and} \\
A(\omega) &= \sum_{j=-\infty}^{\infty}a_j e^{i\omega j}.
\end{aligned}
$$

Then it is easy to see that $2\pi f(\omega)=|A(\omega)|^2$, and

$$
\begin{aligned}
T_n(\omega) &= |R_n(\omega)+A(\omega)J_{\varepsilon,n}(\omega)|^2-|A(\omega)|^2 I_{\varepsilon,n}(\omega) \\
&= R_n(\omega)\bar{A}(\omega)\bar{J}_{\varepsilon,n}(\omega)+\bar{R}_n(\omega)A(\omega)J_{\varepsilon,n}(\omega)+|R_n(\omega)|^2.
\end{aligned}
$$

Now

$$
\begin{aligned}
R_n(\omega) &= J_{x,n}(\omega)-A(\omega)J_{\varepsilon,n}(\omega) \\
&= \frac{1}{\sqrt{n}}\sum_{j=0}^{n-1}\Big(\sum_{t=-\infty}^{\infty}a_t\varepsilon_{j-t}\Big)e^{i\omega j}-\frac{1}{\sqrt{n}}\sum_{t=-\infty}^{\infty}a_t e^{i\omega t}\sum_{j=0}^{n-1}\varepsilon_j e^{i\omega j} \\
&= \frac{1}{\sqrt{n}}\sum_{t=-\infty}^{\infty}a_t e^{i\omega t}\Big[\sum_{j=0}^{n-1}\varepsilon_{j-t}e^{i\omega(j-t)}-\sum_{j=0}^{n-1}\epsilon_j e^{i\omega j}\Big] \\
&= \frac{1}{\sqrt{n}}\sum_{t=-\infty}^{\infty}a_t e^{i\omega t}Z_{n,t}(\omega), \quad \text{say}.
\end{aligned}
$$

Observe that $|Z_{n,0}(\omega)| = 0$ and

$$
|Z_{n,t}(\omega)| \leq
\begin{cases}
|\sum\limits_{j=-t}^{-1} \varepsilon_l e^{i\omega j}| + |\sum\limits_{j=n-t}^{n-1} \varepsilon_j e^{i\omega j}|, & 1 \leq t < n \\[3mm]
|\sum\limits_{j=n}^{n-1-t} \varepsilon_j e^{i\omega j}| + |\sum\limits_{l=0}^{-t-1} \varepsilon_j e^{i\omega j}|, & -n < t \leq -1 \\[3mm]
|\sum\limits_{-t}^{n-1-t} \varepsilon_j e^{i\omega j}| + |\sum\limits_{0}^{n-1} \varepsilon_j e^{i\omega j}|, & |t| \geq n.
\end{cases}
\tag{7.4}
$$

Therefore using (7.3) and (7.4), we get

$$
\begin{aligned}
E(\max_{0 \leq \omega \leq \pi} |R_n(\omega)|) &\leq \frac{2K^{\frac{1}{2}}}{\sqrt{n}} \Big[\sum_{t=1}^{n-1} |a_t| t^{\frac{3}{4}} + \sum_{t=-n+1}^{-1} |a_t| |t|^{\frac{3}{4}} \\
&\quad + \sum_{t=n}^{\infty} |a_t| n^{\frac{3}{4}} + \sum_{t=-\infty}^{-n} |a_t| n^{\frac{3}{4}} \Big] \\
&= \frac{2K^{\frac{1}{2}}}{\sqrt{n}} \Big[\sum_{1 \leq |t| \leq n-1} |a_t| |t|^{\frac{3}{4}} + \sum_{|t| \geq n} |a_t| n^{\frac{3}{4}} \Big] \\
&< 2K^{\frac{1}{2}} n^{-\frac{1}{4}} \Big[\sum_{1 \leq |t| \leq n-1} |a_t| |t|^{\frac{1}{2}} (|t|/n)^{\frac{1}{4}} + \sum_{|t| \geq n} |a_t| |t|^{\frac{1}{2}} \Big] \\
&= o(n^{-1/4}),
\end{aligned}
\tag{7.5}
$$

since the second sum goes to zero as $n \to \infty$ by Assumption 7.1.1, and the first sum is clearly bounded by

$$
\sum_{k(n) < |t| \leq n-1} |t|^{1/2} |a_t| + \{k(n)/n\}^{1/4} \sum_{1 \leq |j| \leq k(n)} |j|^{1/2} |a_j|,
$$

if we choose $k(n)$ such that

$$
\lim_{n \to \infty} \{k(n)/n\} = 0 \quad \text{and} \quad \lim_{n \to \infty} k(n) = \infty.
$$

It is known from Davis and Mikosch (1999) (see Theorem 11.3.2 in Appendix) that under the conditions assumed on $\{\varepsilon_t\}$,

$$
\max_{1 \leq k \leq n} |I_{\varepsilon,n}(\omega_k)| = O_p(\ln n),
$$

and hence

$$
\max_{1 \leq k \leq n} |J_{\varepsilon,n}(\omega_k)| = O_p(\sqrt{\ln n}).
\tag{7.6}
$$

Finally using (7.5) and (7.6),

$$
\max_{1 \leq k < \frac{n}{2}} \left| \frac{I_{x,n}(\omega_k)}{2\pi f(\omega_k)} - I_{\varepsilon,n}(\omega_k) \right|
$$

$$
\leq \frac{1}{2\pi\alpha} \max_{1 \leq k < \frac{n}{2}} |T_n(\omega_k)|
$$

$$
\leq \frac{1}{2\pi\alpha} \Big[2 \max_{0 \leq \omega \leq \pi} |R_n(\omega)| \max_{0 \leq \omega \leq \pi} |A(\omega)| \max_{1 \leq \omega_k < \frac{n}{2}} |J_{\varepsilon,n}(\omega_k)|
$$

$$
+ \{ \max_{0 \leq \omega \leq \pi} |R_n(\omega)| \}^2 \Big]
$$

$$
= o_p(n^{-1/4} \sqrt{\ln n}). \tag{7.7}
$$

This completes the proof. $\qquad \square$

Proof of Theorem 7.2.1. From Lemma 7.2.2, we have

$$
\max_{1 \leq k < \frac{n}{2}} \left| \frac{I_{x,n}(\omega_k)}{2\pi f(\omega_k)} - I_{\varepsilon,n}(\omega_k) \right| = o_p(1), \tag{7.8}
$$

where

$$
I_{x,n}(\omega_k) = \frac{1}{n} \Big| \sum_{t=0}^{n-1} x_t e^{-it\omega_k} \Big|^2 \quad \text{and} \quad I_{\varepsilon,n}(\omega_k) = \frac{1}{n} \Big| \sum_{t=0}^{n-1} \varepsilon_t e^{-it\omega_k} \Big|^2.
$$

Combining this with Theorem 2.1 of Davis and Mikosch (1999) (see Theorem 11.3.2 in Appendix), we have

$$
\max_{1 \leq k < \frac{n}{2}} \frac{I_{x,n}(\omega_k)}{2\pi f(\omega_k)} - \ln q \xrightarrow{\mathcal{D}} \Lambda.
$$

Now proceeding as in the proof of Theorem 5.1.1, we can conclude that

$$
\frac{M(\frac{1}{\sqrt{n}} RC_n, f) - d_q}{c_q} \xrightarrow{\mathcal{D}} \Lambda.
$$

$\qquad \square$

Remark 7.2.1. If we include λ_0 in the maximum and define

$$
\overline{M}(\frac{1}{\sqrt{n}} RC_n, f) = \max_{0 \leq k < n/2} \frac{|\lambda_k|}{\sqrt{2\pi f(\omega_k)}},
$$

then different limits may appear depending on the mean μ of the process $\{x_n\}$.

If the mean μ of ε_0 is 0 then by Theorem 7.1.2 of Brockwell and Davis (2006) (see Theorem 11.3.1 in Appendix), it follows that $\frac{\lambda_0}{\sqrt{2\pi f(0)}} \xrightarrow{\mathcal{D}} N(0,1)$. So by arguments similar to Theorem 5.1.1 we have

$$
\frac{\overline{M}(\frac{1}{\sqrt{n}} RC_n, f) - d_q}{c_q} \xrightarrow{\mathcal{D}} \Lambda.
$$

If, on the other hand $\mu \neq 0$, then

$$\overline{M}(\frac{1}{\sqrt{n}}RC_n, f) - |\mu|\sqrt{n} \xrightarrow{D} N(0,1).$$

7.3 Symmetric circulant

Define $M(\cdot, f)$ for the symmetric circulant matrix as was done for the reverse circulant matrix:

$$M(\frac{1}{\sqrt{n}}SC_n, f) = \max_{1 \leq k < \frac{n}{2}} \frac{|\lambda_k|}{\sqrt{2\pi f(\omega_k)}},$$

where λ_k are the eigenvalues of $\frac{1}{\sqrt{n}}SC_n$ as defined in Section 1.2 of Chapter 1. Under the additional restriction of $a_j = a_{-j}$, for all j, we find the limiting distribution of $M(\frac{1}{\sqrt{n}}SC_n, f)$.

Theorem 7.3.1 (Bose et al. (2011c)). Suppose $\{x_n\}$ is the two-sided moving average process defined in (7.1) with $a_j = a_{-j}$, and satisfies Assumption 7.1.1. Then

$$\frac{M(\frac{1}{\sqrt{n}}SC_n, f) - b_q - a_q \ln 2}{a_q} \xrightarrow{D} \Lambda, \tag{7.9}$$

where $q = q(n) = \lfloor \frac{n}{2} \rfloor \approx \frac{n}{2}$, and a_q and b_q are as in (5.4) of Chapter 5.

As in Remark 7.2.1, if λ_0 is included in the definition of $M(\frac{1}{\sqrt{n}}SC_n, f)$, then the result changes only when $\mu \neq 0$.

Before providing the proof of Theorem 7.3.1, we prove the following result which is similar to Lemma 7.2.2. It will be used in the proof of Theorem 7.3.1.

Lemma 7.3.2. Let $\{x_n\}$ be the two-sided moving average process (7.1) where $E(\varepsilon_i) = 0$, $E(\varepsilon_i^2) = 1$,

$$\sum_{j=-\infty}^{\infty} |a_j||j|^{1/2} < \infty, \text{ and } f(\omega) > 0 \text{ for all } \omega \in [0, 2\pi].$$

Then we have

$$\max_{1 \leq k \leq \lfloor \frac{n}{2} \rfloor} \left| \frac{\lambda_k}{\sqrt{2\pi f(\omega_k)}} - 2\frac{A_k}{\sqrt{n}} \sum_{t=1}^{\lfloor \frac{n}{2} \rfloor} \varepsilon_t \cos(\frac{2\pi kt}{n}) + 2\frac{B_k}{\sqrt{n}} \sum_{t=1}^{\lfloor \frac{n}{2} \rfloor} \varepsilon_t \sin(\frac{2\pi kt}{n}) \right| = o_p(n^{-\frac{1}{4}}),$$

where

$$\sqrt{2\pi f(\omega_k)}A_k = \sum_{j=-\infty}^{\infty} a_j \cos(\frac{2\pi kj}{n}), \text{ and}$$

$$\sqrt{2\pi f(\omega_k)}B_k = \sum_{j=-\infty}^{\infty} a_j \sin(\frac{2\pi kj}{n}).$$

Proof. First observe that $\min_{\omega \in [0, 2\pi]} f(\omega) > \alpha > 0$. Consider $n = 2m + 1$ for simplicity. For $n = 2m$ the calculations are similar.

$$\frac{\lambda_k}{\sqrt{2\pi f(\omega_k)}} - 2\frac{A_k}{\sqrt{n}} \sum_{t=1}^{m} \varepsilon_t \cos(\frac{2\pi kt}{n}) + 2\frac{B_k}{\sqrt{n}} \sum_{t=1}^{m} \varepsilon_t \sin(\frac{2\pi kt}{n}) = Y_{n,k},$$

where

$$Y_{n,k} = \frac{1}{\sqrt{n}\sqrt{2\pi f(\omega_k)}} \sum_{j=-\infty}^{\infty} a_j \left[\cos\frac{2\pi kj}{n} U_{k,j} - \sin\frac{2\pi kj}{n} V_{k,j} \right],$$

$$U_{k,j} = \sum_{t=1}^{m} \left[\varepsilon_{t-j} \cos\frac{2\pi k(t-j)}{n} - \varepsilon_t \cos\frac{2\pi kt}{n} \right], \quad \text{and}$$

$$V_{k,j} = \sum_{t=1}^{m} \left[\varepsilon_{t-j} \sin\frac{2\pi k(t-j)}{n} - \varepsilon_t \sin\frac{2\pi kt}{n} \right].$$

Now using (7.2), we get

$$\mathrm{E}\{\max_k U_{k,j}^2\} \leq \begin{cases} 4K|j|^{3/2} & \text{if} \quad |j| < m \\ 4Km^{3/2} & \text{if} \quad |j| \geq m, \text{ and} \end{cases} \tag{7.10}$$

$$\mathrm{E}\{\max_k V_{k,j}^2\} \leq \begin{cases} 4K|j|^{3/2} & \text{if} \quad |j| < m \\ 4Km^{3/2} & \text{if} \quad |j| \geq m. \end{cases} \tag{7.11}$$

Therefore

$$\begin{aligned} \mathrm{E}\{\max_k |Y_{n,k}|\} &\leq \frac{1}{\sqrt{2\pi\alpha}} \frac{1}{\sqrt{n}} \sum_{j=-\infty}^{\infty} |a_j| \left[\mathrm{E}\{\max_k |U_{k,j}|\} + \mathrm{E}\{\max_k |V_{k,j}|\} \right] \\ &\leq \frac{2K^{1/2}}{\sqrt{2\pi\alpha}} \frac{1}{\sqrt{n}} \left[\sum_{|j|<m} |a_j||j|^{3/4} + \sum_{|j|\geq m} |a_j|m^{3/4} \right] \\ &\leq \frac{2K^{1/2}}{\sqrt{2\pi\alpha}} \frac{1}{n^{1/4}} \left[\sum_{|j|<m} |j|^{1/2}|a_j|(j/n)^{1/4} + \sum_{|j|\geq m} j^{1/2}|a_j| \right] \\ &= o(n^{-1/4}), \end{aligned}$$

by the same argument that was used to obtain (7.5).

□

Proof of Theorem 7.3.1. Note that if $a_j = a_{-j}$ then in Lemma 7.3.2, $B_k = 0$ and hence from the same lemma, it is easy to see that

$$\max_{1 \leq k \leq \lfloor \frac{n}{2} \rfloor} \left| \frac{\lambda_k}{\sqrt{2\pi f(\omega_k)}} - \lambda_{k,\varepsilon} \right| = o_p(n^{-1/4}), \tag{7.12}$$

where $\lambda_{k,\varepsilon}$ denote the eigenvalues of the symmetric circulant matrix with $\{x_i\}$ replaced by $\{\varepsilon_i\}$. Combining this with Theorem 5.2.5(b), we have the result (7.9).

□

In Theorem 7.3.1, we assumed that $a_j = a_{-j}$. Now we focus on the case where a_j is not necessarily equal to a_{-j}. For reasons to be discussed later (see Remark 7.3.2), in this case we will deal with the maximum over two different subsets L_n^1 and L_n^2 (see (7.14)) of $\{1 \le k \le \lfloor \frac{n}{2} \rfloor\}$ separately. We first define some notation to state our future results. For $0 < \delta_1 < 1/2$, define

$$p_n = 1 - \frac{1}{n^{1/2+\delta_1}}, \quad L_n = \{k : 1 \le k \le \lfloor np_n/2 \rfloor\}, \tag{7.13}$$

$$L_n^1 = \{k \in L_n : k \text{ is even}\}, \quad \text{and} \quad L_n^2 = \{k \in L_n : k \text{ is odd}\}. \tag{7.14}$$

Theorem 7.3.3 (Bose et al. (2011c)). Let SC_n be the symmetric circulant matrix with input sequence $\{x_n\}$ defined in (7.1) and which satisfies Assumption 7.1.1. Then $\{\lambda_{k,x}\}$, the eigenvalues of $\frac{1}{\sqrt{n}}SC_n$, satisfy

$$\frac{\max_{k \in L_n^1} \lambda_{k,x} - b_q}{a_q} \xrightarrow{\mathcal{D}} \Lambda, \tag{7.15}$$

and

$$\frac{\max_{k \in L_n^2} \lambda_{k,x} - b_q}{a_q} \xrightarrow{\mathcal{D}} \Lambda, \tag{7.16}$$

where $q = q(n) = \lfloor \frac{n}{4} \rfloor$, and a_q and b_q are as in (5.4).

We first prove a result similar to Theorem 7.3.3 with normal entries in Lemma 7.3.5. Then we prove this theorem. We define some notation which will be used in further developments. Let

$$\begin{aligned} \sigma_k^2 &= 1 + \frac{A_k B_k}{n} \tan\left(\frac{\pi k}{n}\right), \\ \nu_{k,k'} &= \frac{D_{k,k'}}{n} \tan\frac{\pi(k+k')}{2n} + \frac{E_{k,k'}}{n} \tan\frac{\pi(k'-k)}{2n}, \\ D_{k,k'} &= A_k B_{k'} + A_{k'} B_k, \quad \text{and} \quad E_{k,k'} = A_{k'} B_k - A_k B_{k'}, \end{aligned} \tag{7.17}$$

where A_k, B_k are as in Lemma 7.3.2.

The following result from Dai and Mukherjea (2001) (Theorem 2.1) is an analogue of Mill's ratio in higher dimension.

Theorem 7.3.4 (Dai and Mukherjea (2001)). Let $(X_1, X_2, \ldots, X_n)$ be multivariate normal with zero means and a positive definite covariance matrix Σ. Let $\tilde{\sigma}_1 \ge \tilde{\sigma}_2 \ge \cdots \ge \tilde{\sigma}_n$ denote the variances and let $I(t) = \mathrm{P}(X_i \ge t, 1 \le i \le n)$. If $\alpha = (\alpha_1, \alpha_2, \ldots, \alpha_n) = \vec{1}\Sigma^{-1}$, where $\vec{1} = (1, 1, \ldots, 1)$ with $\alpha_i > 0$, then

$$I(t) \approx \frac{1}{(\sqrt{2\pi})^n \sqrt{|\Sigma|}(\prod_{i=1}^n \alpha_i)t^n} \exp\left(-\frac{1}{2}t^2 \vec{1}\Sigma^{-1}\vec{1}^T\right).$$

We first look at the special case where $\{\varepsilon\}$ are standard normal random variables.

Lemma 7.3.5. Let $\{N_i\}$ be i.i.d. $N(0,1)$ and let

$$\lambda_{k,N} = \frac{\sqrt{2}A_k N_0}{\sqrt{n}} + \frac{1}{\sqrt{n}}\sum_{t=1}^{\lfloor \frac{n}{2} \rfloor} N_t\left(2A_k \cos(\frac{2\pi kt}{n}) - 2B_k \sin(\frac{2\pi kt}{n})\right).$$

Then

$$\frac{\max_{k \in L_n^1} \lambda_{k,N} - b_q}{a_q} \xrightarrow{\mathcal{D}} \Lambda, \qquad (7.18)$$

and

$$\frac{\max_{k \in L_n^2} \lambda_{k,N} - b_q}{a_q} \xrightarrow{\mathcal{D}} \Lambda, \qquad (7.19)$$

where $q = q(n) = \lfloor \frac{n}{4} \rfloor$, and a_n and b_n are as in (5.4).
   In particular,

$$\frac{\max_{1 \leq k \leq \lfloor \frac{n}{2} \rfloor} \lambda_{k,N}}{\sqrt{2\ln n}} \xrightarrow{\mathcal{P}} 1. \qquad (7.20)$$

Proof. We shall prove only (7.18). Proof of (7.19) is similar. Finally, using these two results we shall prove (7.20).

Proof of (7.18): Consider the case $n = 2m+1$. For $n = 2m$, the calculations will be similar with minor changes. First observe that $\mathrm{Var}(\lambda_{k,N}) = \sigma_k^2$, and for $k' > k$ we have $\mathrm{Cov}(\lambda_{k,N}, \lambda_{k',N}) = \nu_{k,k'}$, where σ_k and $\nu_{k,k'}$ are defined in (7.17). Let $x_q = a_q x + b_q \approx \sqrt{2\ln q}$. By Bonferroni inequalities we have for $j > 1$,

$$\sum_{d=1}^{2j}(-1)^{d-1}\tilde{B}_d \leq P(\max_{k \in L_n^1} \lambda_{k,N} > x_q) \leq \sum_{d=1}^{2j-1}(-1)^{d-1}\tilde{B}_d,$$

where

$$\tilde{B}_d = \sum_{\substack{i_1,i_2,\ldots,i_d \in L_n^1 \\ \text{all distinct}}} P(\lambda_{i_1,N} > x_q,\ldots \lambda_{i_d,N} > x_q).$$

Observe that by the choice of p_n as given in (7.13), we have

$$\frac{1}{n}\tan(\frac{\pi p_n}{2}) \approx \frac{2n^{1/2+\delta_1}}{\pi n} \to 0.$$

Hence for some $\epsilon > 0$, for large n we have $1 - \epsilon < \sigma_k^2 < 1 + \epsilon$, and for any $k, k' \in L_n^1$ (or L_n^2) we have $|\nu_{k,k'}| \to 0$ as $n \to \infty$. We shall use this simple observation very frequently in the proof. Next we make the following claim.

Claim:

$$\sum_{\substack{i_i,i_2,\ldots,i_d \in L_n^1 \\ \text{all distinct}}} \mathbb{P}(\lambda_{i_1,N} > x_q,\ldots \lambda_{i_d,N} > x_q) \approx \frac{q^d \exp(-\frac{x_q^2 d}{2})}{d! x_q^d (\sqrt{2\pi})^d}, \quad \text{for } d \geq 1. \quad (7.21)$$

To avoid notational complications, we establish the above claim for $d = 1$ and $d = 2$, and indicate what changes are necessary for a higher dimension.

d=1: Using the fact that $\dfrac{\sigma_k^2}{x_q^2} \to 0$, and for $x > 0$,

$$\left(1 - \frac{1}{x^2}\right)\frac{\exp(-x^2/2)}{\sqrt{2\pi}x} \leq \mathrm{P}(N(0,1) > x) \leq \frac{\exp(-x^2/2)}{\sqrt{2\pi}x},$$

it easily follows that

$$\sum_{k \in L_n^1} \mathrm{P}(N(0,1) > x_q/\sigma_k) \approx \sum_{k \in L_n^1} \frac{\sigma_k}{\sqrt{2\pi}x_q}\exp(-\frac{x_q^2}{2\sigma_k^2}).$$

Observe that

$$\frac{\sum_{k \in L_n^1}\frac{\sigma_k}{\sqrt{2\pi}x_q}\exp(-\frac{x_q^2}{2\sigma_k^2})}{\frac{qp_n}{\sqrt{2\pi}x_q}\exp(-\frac{x_q^2}{2})} = \frac{1}{qp_n}\sum_{k \in L_n^1}\sigma_k\exp(-\frac{x_q^2}{2}(\frac{1}{\sigma_k^2} - 1))$$

$$= \frac{1}{qp_n}\sum_{k \in L_n^1}\sigma_k\exp(-\frac{x_q^2}{2\sigma_k^2}\frac{A_kB_k}{n}\tan(\frac{\pi k}{n})).$$

The last term above goes to 1. Since $p_n \approx 1$, the claim is proved for $d = 1$.

d=2: We shall use Lemma 7.3.4 for this case. Without loss of generality assume that $\sigma_k^2 > \sigma_{k'}^2$. Let $\alpha = (\alpha_1, \alpha_2)$, where $\alpha = \vec{1}V^{-1}$ and

$$V = \begin{bmatrix} \sigma_k^2 & \nu_{k,k'} \\ \nu_{k,k'} & \sigma_{k'}^2 \end{bmatrix}.$$

Hence $(\alpha_1, \alpha_2) = \left(\frac{\sigma_{k'}^2 - \nu_{k,k'}}{|V|}, \frac{\sigma_k^2 - \nu_{k,k'}}{|V|}\right)$, where $|V|$ is the determinant of V. For any $0 < \epsilon < 1$, it easily follows that $\alpha_i > \frac{1-\epsilon}{|V|}$ for large n and for $i = 1, 2$. Hence, from Lemma 7.3.4 it follows that as $n \to \infty$,

$$\sum_{k,k' \in L_n^1} \mathrm{P}(\lambda_{k,N} > x_q, \lambda_{k',N} > x_q) \approx \sum_{k,k' \in L_n^1}\frac{1}{2\pi\sqrt{|V|}}\frac{\exp(-\frac{1}{2}x_q^2\vec{1}V^{-1}\vec{1}^T)}{\alpha_1\alpha_2x_q^2}.$$

Now denote

$$\psi_{k,k'} = \frac{1}{|V|}\Big[-\frac{A_kB_k}{n}\tan(\frac{\pi k}{n}) - \frac{A_{k'}B_{k'}}{n}\tan(\frac{\pi k'}{n})$$

$$+\frac{A_kB_k}{n}\tan(\frac{\pi k}{n})\frac{A_{k'}B_{k'}}{n}\tan(\frac{\pi k'}{n}) - 2\nu_{k,k'} + 2\nu_{k,k'}^2\Big],$$

and observe that

$$|x_q^2\psi_{k,k'}| \leq C\frac{x_q^2}{n}\tan(\frac{\pi p_n}{2}) \to 0 \quad \text{as} \quad n \to \infty.$$

$$\frac{\sum_{k,k'\in L_n^1}\frac{1}{2\pi\sqrt{|V|\alpha_1\alpha_2 x_q^2}}\exp(-\frac{1}{2}x_q^2\vec{1}V^{-1}\vec{1}^T)}{\frac{q^2\exp(-x_q^2)}{2!x_q^2 2\pi}}$$

$$=\frac{2}{q^2}\sum_{k,k'\in L_n^1}\frac{1}{\sqrt{|V|\alpha_1\alpha_2}}\exp\left(-\frac{1}{2}x_q^2(\alpha_1+\alpha_2)+x_q^2\right)$$

$$=\frac{2}{q^2}\sum_{k,k'\in L_n^1}\frac{|V|^{3/2}}{(\sigma_{k'}^2-\nu_{k,k'})(\sigma_k^2-\nu_{k,k'})}\exp\left(-\frac{x_q^2}{2}(\alpha_1+\alpha_2-2)\right)$$

$$\leq\frac{2}{q^2}\sum_{k,k'\in L_n^1}\frac{|V|^{3/2}}{(1-\epsilon)^2}\exp(-\frac{x_q^2}{2}\psi_{k,k'})\to 1 \text{ as } n\to\infty \text{ and as } \epsilon\to 0.$$

A lower bound can be obtained similarly to verify the claim for $d=2$.

d > 2 : The probability inside the sum in claim (7.21) is $P(N(0,V_n)\in E_n)$ where $E_n=\{(y_1,y_2,\ldots,y_d)^T : y_i > x_q, i=1,2\ldots,d\}$, and V_n denotes the covariance matrix $\{V_n(s,t)\}_{s,t=1}^d$ with $V_n(s,s)=\sigma_{i_s}^2$ and for $s\neq t$, $V_n(s,t)=\nu_{i_s i_t}$, where $\sigma_{i_s},\nu_{i_s i_t}$ are as in (7.17). Without loss of generality assume that $\sigma_{i_1}\geq\sigma_{i_2}\geq\cdots\geq\sigma_{i_d}$, since we can always permute the original vector to achieve this, and the covariance matrix changes accordingly. Note that

$$\|V_n-I_d\|_\infty\to 0 \text{ as } n\to\infty,$$

where $\|A\|_\infty=\max|a_{i,j}|$. As $V_n^{-1}=\sum_{j=0}^\infty(I_d-V_n)^j$, we have $\alpha=\vec{1}+\sum_{j=1}^\infty\vec{1}(I_d-V_n)^j$. Now since $\|I_d-V_n\|_\infty\to 0$, we have $\|(I_d-V_n)^j\|_\infty\to 0$ and hence elements of $(I_d-V_n)^j$ go to zero for all j. So we get that $\alpha_i\in(1-\epsilon,1+\epsilon)$ for $i=1,2,\ldots,d$ and $0<\epsilon<1$. Hence, we can again apply Lemma 7.3.4. For further calculations it is enough to observe that for $x=(x_1,x_2,\ldots,x_d)\neq 0$,

$$\frac{xV_nx^T}{|x|^2}=1+\frac{1}{|x|^2}\sum_{k=1}^d x_k^2 A_{i_k}B_{i_k}\frac{1}{n}\tan(\frac{\pi i_k}{n})+\frac{1}{|x|^2}\sum_{1\leq k\neq k'\leq d}x_k x_{k'}\nu_{i_k,i_{k'}}.$$

Since the last two terms go to zero, given any $\epsilon>0$, we get for large n,

$$1-\epsilon\leq\lambda_{min}(V_n)\leq\lambda_{max}(V_n)\leq 1+\epsilon,$$

where $\lambda_{min}(V_n)$ and $\lambda_{max}(V_n)$ denote the minimum and the maximum eigenvalues of V_n. The rest of the calculation is similar to the $d=2$ case. This proves the claim completely.

Back to the proof of (7.18). Using the fact that a_n and b_n are normalizing constants for the maxima of i.i.d. standard normal variables, it follows that

$$\frac{q^d\exp(-\frac{x_q^2 d}{2})}{d!x_q^d(\sqrt{2\pi})^d}\approx\frac{1}{d!}\exp(-dx).$$

So from the Bonferroni inequalities, and observing $\exp(-\exp(-x)) = \sum_{d=0}^{\infty} \frac{(-1)^d}{d!} \exp(-dx)$, it follows that

$$P(\max_{k \in L_n^1} \lambda_{k,N} > x_q) \to \exp(-\exp(-x)),$$

proving (7.18) completely.

Proof of (7.20): We first observe that

$$\sum_{k=np_n/2}^{n/2} P(N(0,1) > x_q/\sigma_k) \le \frac{n}{2}(1-p_n)\,P(N(0,1) > \frac{x_q}{\sqrt{2}}),$$

since $\sigma_k^2 \le 2$ for $k \le n/2$. Expanding the expressions for a_n and b_n, we get

$$\frac{x_q^2}{4} = \frac{1}{4}(a_q x + b_q)^2 = o(1) + \frac{\ln q}{2} - \frac{1}{4}\ln(4\pi \ln q) + \frac{x}{2}.$$

Now

$$\frac{n(1-p_n)}{2}\,P(N(0,2) > x_q) \le C\frac{n(1-p_n)}{2}\frac{\exp(-\frac{x_q^2}{4})}{x_q}$$

$$\approx C n^{-1/2}\frac{n(1-p_n)}{2\sqrt{\ln q}}$$

$$\approx C\frac{1}{n^{\delta_1}\sqrt{\ln q}} \to 0 \text{ as } n \to \infty.$$

Breaking up the set $L_1 = \{k : 1 \le k \le \lfloor \frac{n}{2} \rfloor$ and k is even$\}$ into L_n^1 and $\tilde{L}_n^1 = \{k : \lfloor np_n/2 \rfloor < k \le \lfloor \frac{n}{2} \rfloor$ and k is even$\}$, we get

$$P(\max_{k \in L_1} \lambda_{k,N} > x_q) = P(\max(\max_{k \in L_n^1} \lambda_{k,N}, \max_{k \in \tilde{L}_n^1} \lambda_{k,N}) > x_q)$$

$$\le P(\max_{k \in L_n^1} \lambda_{k,N} > x_q) + P(\max_{k \in \tilde{L}_n^1} \lambda_{k,N} > x_q)$$

$$\le P(\max_{k \in L_n^1} \lambda_{k,N} > x_q) + \sum_{t=\lfloor np_n/2 \rfloor}^{\lfloor \frac{n}{2} \rfloor} P(N(0,\sigma_k^2) > x_q)$$

$$= P(\max_{k \in L_n^1} \lambda_{k,N} > x_q) + o(1).$$

Hence the upper bound is obtained. The lower bound easily follows from (7.18). Similar calculations for the set $L_2 = \{k : 1 \le k < \lfloor \frac{n}{2} \rfloor$ and k is odd$\}$ can be done. To complete the proof, observe that

$$P\left(\frac{\max_{1 \le k < \lfloor \frac{n}{2} \rfloor} \lambda_{k,N}}{\sqrt{2\ln n}} > 1 + \epsilon\right) \le P\left(\frac{\max_{k \in L_1} \lambda_{k,N}}{\sqrt{2\ln n}} > 1 + \epsilon\right)$$

$$+ P\left(\frac{\max_{k \in L_2} \lambda_{k,N}}{\sqrt{2\ln n}} > 1 + \epsilon\right),$$

and both these probabilities go to zero as $n \to \infty$. $\qquad\square$

Remark 7.3.1. By calculations similar to those given above, it can be shown that for $\sigma^2 = n^{-c}$ where $c > 0$,

$$\sum_{\substack{i_1,\ldots,i_d \in L_n^1 \\ \text{all distinct}}} \mathrm{P}((1+\sigma^2)^{1/2}\lambda_{i_1,N} > x_q, \ldots, (1+\sigma^2)^{1/2}\lambda_{i_d,N} > x_q) \leq \frac{C^d}{d!}, \quad (7.22)$$

for some constant $K > 0$. This will be used in the proof of Theorem 7.3.3.

Now we prove Theorem 7.3.3 using Lemma 7.3.5.

Proof of Theorem 7.3.3. We shall prove only (7.15). Proof of (7.16) is similar. Again for simplicity we assume that $n = 2m+1$. We break the proof into four steps.

Step 1: Truncation: Define

$$\tilde{\varepsilon}_t = \varepsilon_t \mathbb{I}(|\varepsilon_t| \leq n^{\frac{1}{2+\delta}}), \quad \bar{\varepsilon}_t = \tilde{\varepsilon}_t - E\tilde{\varepsilon}_t, \quad \tilde{x}_t = \sum_{j=-\infty}^{\infty} a_j \tilde{\varepsilon}_{t-j}, \quad \bar{x}_t = \sum_{j=-\infty}^{\infty} a_j \bar{\varepsilon}_{t-j},$$

$$\lambda_{k,\tilde{x}} = \frac{1}{\sqrt{n}}[\tilde{x}_0 + 2\sum_{t=1}^{m} \tilde{x}_t \cos\frac{2\pi kt}{n}], \quad \lambda_{k,\bar{x}} = \frac{1}{\sqrt{n}}[\bar{x}_0 + 2\sum_{t=1}^{m} \bar{x}_t \cos\frac{2\pi kt}{n}].$$

Claim: To prove (7.15), it is enough to show that

$$\frac{\max_{k \in L_n^1} \lambda_{k,\varepsilon} - b_q}{a_q} \xrightarrow{\mathcal{D}} \Lambda, \quad (7.23)$$

where

$$\lambda_{k,\varepsilon} = \frac{\sqrt{2}A_k \bar{\varepsilon}_0}{\sqrt{n}} + \frac{2A_k}{\sqrt{n}} \sum_{t=1}^{m} \bar{\varepsilon}_t \cos(\frac{2\pi kt}{n}) - \frac{2B_k}{\sqrt{n}} \sum_{t=1}^{m} \bar{\varepsilon}_t \sin(\frac{2\pi kt}{n}).$$

To prove the claim first note that

$$\lambda_{k,\bar{x}} = \lambda_{k,\tilde{x}}. \quad (7.24)$$

Choose η such that $(\frac{1}{2} - \frac{1}{2+\delta} - \eta) > 0$ and observe that

$$n^\eta E[\max_{1 \leq k \leq \lfloor \frac{n}{2} \rfloor} |\lambda_{k,\bar{x}} - \lambda_{k,x}|] = n^\eta E[\max_{1 \leq k \leq \lfloor \frac{n}{2} \rfloor} |\lambda_{k,\tilde{x}} - \lambda_{k,x}|]$$

$$\leq \frac{2}{n^{1/2-\eta}} \sum_{t=0}^{m} \sum_{j=-\infty}^{\infty} |a_j| E(|\varepsilon_{t-j}|\mathbb{I}(|\varepsilon_{t-j}| > n^{\frac{1}{2+\delta}}))$$

$$\leq \frac{2}{n^{1/2-\eta}} \sum_{t=0}^{m} \sum_{j=-\infty}^{\infty} |a_j| [n^{\frac{1}{2+\delta}} \mathrm{P}(|\varepsilon_{t-j}| > n^{\frac{1}{2+\delta}})$$

$$+ \int_{n^{\frac{1}{2+\delta}}}^{\infty} \mathrm{P}(|\varepsilon_{t-j}| > u)du]$$

$$= I_1 + I_2, \text{ say,}$$

and

$$I_1 = \frac{2}{n^{1/2-\eta}} \sum_{t=0}^{m} \sum_{j=-\infty}^{\infty} |a_j| n^{\frac{1}{2+\delta}} P(|\varepsilon_{t-j}| > n^{\frac{1}{2+\delta}})$$

$$\leq \frac{2}{n^{1/2-\eta}} \sum_{t=0}^{m} \sum_{j=-\infty}^{\infty} |a_j| n^{\frac{1}{2+\delta}} \frac{1}{n} E(|\varepsilon_{t-j}|^{2+\delta})$$

$$\leq \frac{E(|\varepsilon_0|^{2+\delta})}{n^{\frac{1}{2} - \frac{1}{2+\delta} - \eta}} \sum_{j=-\infty}^{\infty} |a_j| \to 0, \text{ as } n \to \infty.$$

Similarly, $I_2 \to 0$ as $n \to \infty$. Hence

$$\max_{1 \leq k \leq \lfloor \frac{n}{2} \rfloor} |\lambda_{k,\bar{x}} - \lambda_{k,x}| = o_p\left(n^{-\eta}\right). \tag{7.25}$$

Also from Lemma 7.3.2 we have

$$\max_{k \in L_n^1} \left| \frac{\lambda_{k,\bar{x}}}{a_q \sqrt{2\pi f(\omega_k)}} - \frac{2A_k}{\sqrt{n}a_q} \sum_{t=1}^{m} \bar{\varepsilon}_t \cos(\frac{2\pi kt}{n}) + \frac{2B_k}{\sqrt{n}a_q} \sum_{t=1}^{m} \bar{\varepsilon}_t \sin(\frac{2\pi kt}{n}) \right|$$

$$= o_p(\frac{\sqrt{\ln n}}{n^{\delta_1}}). \tag{7.26}$$

Now (7.25) and (7.26) prove the claim completely.

Step 2: Normal Approximation: This is an intermediate step to approximate $\lambda_{k,\varepsilon}$ by $\lambda_{k,N}$, where $\lambda_{k,N}$ is defined in Lemma 7.3.5. Define

$$\lambda_{k,\epsilon+\sigma N} = \frac{\sqrt{2}A_k}{\sqrt{n}}(\bar{\varepsilon}_0 + \sigma N_0) + \frac{2A_k}{\sqrt{n}} \sum_{t=1}^{m} (\bar{\varepsilon}_t + \sigma N_t) \cos(\frac{2\pi kt}{n})$$

$$- \frac{2B_k}{\sqrt{n}} \sum_{t=1}^{m} (\bar{\varepsilon}_t + \sigma N_t) \sin(\frac{2\pi kt}{n}).$$

Claim:

$$\left| P(\max_{k \in L_n^1} \lambda_{k,\epsilon+\sigma N} > x_q) - P(\max_{k \in L_n^1} (1+\sigma^2)^{1/2} \lambda_{k,N} > x_q) \right| \to 0, \tag{7.27}$$

where $\lambda_{k,N}$ is defined in Lemma 7.3.5.

Proof of this claim is similar to the proof of Lemma 5.2.3. It uses Lemma 5.2.2. We omit the details.

Step 3: In this step we shall prove (7.23). First observe the following:

$$\lim_{n \to \infty} P(\max_{k \in L_n^1} \lambda_{k,\varepsilon+\sigma N} > x_q) = \Lambda(x). \tag{7.28}$$

Proof of this observation is similar to Step 3 of the proof of Theorem 5.2.4. Here we skip the details. Now note that

$$\left| \frac{\max_{k \in L_n^1} \lambda_{k,\varepsilon+\sigma N} - b_q}{a_q} - \frac{\max_{k \in L_n^1} \lambda_{k,\varepsilon} - b_q}{a_q} \right| \leq \frac{\sigma \max_{k \in L_n^1} \lambda_{k,N}}{a_q} \xrightarrow{\mathcal{P}} 0.$$

Now using (7.28) it follows that

$$\frac{\max_{k \in L_n^1} \lambda_{k,\varepsilon} - b_q}{a_q} \xrightarrow{\mathcal{D}} \Lambda.$$

This completes the proof of Step 4, and hence of (7.23). As a consequence (7.15) is proven completely. This completes the proof of the theorem. □

Theorem 7.3.6 (Bose et al. (2011c)). *If* $\{\lambda_{k,x}\}$ *are the eigenvalues of* $\frac{1}{\sqrt{n}} SC_n$ *then under the assumptions of Theorem 7.3.3,*

$$\frac{\max_{1 \le k \le \lfloor \frac{n}{2} \rfloor} \frac{\lambda_{k,x}}{\sqrt{2\pi f(w_k)}}}{\sqrt{\ln n}} \xrightarrow{\mathcal{P}} 1, \quad \text{where } w_k = \frac{2\pi k}{n}.$$

Proof. As before we assume that $n = 2m + 1$. It is now easy to see from the truncation part of Theorem 7.3.3 and Lemma 7.3.2 that it is enough to show that,

$$\frac{\max_{1 \le k \le \lfloor \frac{n}{2} \rfloor} \lambda_{k,\varepsilon}}{\sqrt{\ln n}} \xrightarrow{\mathcal{P}} 1,$$

where

$$\lambda_{k,\varepsilon} = \frac{\sqrt{2} A_k \bar{\varepsilon}_0}{\sqrt{n}} + \frac{2A_k}{\sqrt{n}} \sum_{t=1}^{m} \bar{\varepsilon}_t \cos(\frac{2\pi k t}{n}) - \frac{2B_k}{\sqrt{n}} \sum_{t=1}^{m} \bar{\varepsilon}_t \sin(\frac{2\pi k t}{n}),$$

and $\bar{\varepsilon}_t = \varepsilon_t \mathbb{I}(|\varepsilon_t| \le n^{1/s}) - E[\varepsilon_t \mathbb{I}(|\varepsilon_t| \le n^{1/s})]$. The steps are the same as the steps used to prove (7.20) in Lemma 7.3.5. Observe from there that, to complete the proof, it is enough to show that

$$\sum_{k=\lfloor np_n/2 \rfloor+1}^{\lfloor \frac{n}{2} \rfloor} P(\lambda_{k,\varepsilon} > x_q) \to 0 \text{ as } n \to \infty. \tag{7.29}$$

Let

$$m = \lfloor \frac{n}{2} \rfloor, \ v_1(0) = \sqrt{2} A_k, \text{ and } v_1(t) = 2A_k \cos(\frac{2\pi k t}{n}) - 2B_k \sin(\frac{2\pi k t}{n}).$$

Since $\{\bar{\varepsilon}_t v_1(t)\}$ is a sequence of bounded independent mean zero random variables, by applying Bernstein's inequality (see Section 11.2 of Appendix) we get

$$P(\frac{1}{\sqrt{m}} \sum_{t=0}^{m} \bar{\varepsilon}_t v_1(t) > x_q) \le P(|\sum_{t=0}^{m} \bar{\varepsilon}_t v_1(t)| > \sqrt{m} x_q)$$

$$= P(|\sum_{t=0}^{m} \bar{\varepsilon}_t v_1(t)| > m \frac{x_q}{\sqrt{m}})$$

$$\le 2 \exp \left(- \frac{m x_q^2}{2 \sum_{t=0}^{m} \text{Var}(\varepsilon_t v_1(t)) + \frac{2}{3} C n^{1/s} m \frac{x_q}{\sqrt{m}}} \right).$$

Now observe that

$$D \; := \; \frac{m x_q^2}{2 \sum_{t=0}^{m} \mathrm{Var}(\varepsilon_t v_1(t)) + \frac{2}{3} C n^{1/s} m \frac{x_q}{\sqrt{m}}}$$

$$\geq \; \frac{x_q^2}{4 \frac{1}{n} \sum_{t=0}^{m} \mathrm{Var}(\varepsilon_t v_1(t)) + \frac{4}{3} C n^{1/s-1/2} x_q}$$

$$= \; \frac{x_q^2}{4(1 + \frac{A_k B_k}{n} \tan \frac{\pi k}{n}) + \frac{4}{3} \frac{C x_q}{n^{1/2-1/s}}}$$

$$\geq \; \frac{x_q^2}{4(1 + \frac{2}{\pi}) + o(1)} \geq \frac{x_q^2}{8}.$$

Therefore

$$P(|\sum_{t=0}^{m} \bar{\varepsilon}_t v_1(t)| > \sqrt{m} x_q) \leq 2 \exp(-\frac{x_q^2}{8}),$$

and hence

$$\sum_{t=\lfloor n p_n/2 \rfloor}^{\lfloor \frac{n}{2} \rfloor} P(\frac{1}{\sqrt{n}} |\sum_{t=0}^{m} \bar{\varepsilon}_t v_1(t)| > x_q) \leq n(1 - p_n) \exp(-\frac{x_q^2}{4})$$

$$\leq \frac{C}{n^{\delta_1} (\ln n)^{1/4}} \to 0.$$

This completes the proof of (7.29), and hence the proof of the theorem. □

Remark 7.3.2. In Theorem 7.3.3 we were unable to consider the convergence over $L_n^1 \cup L_n^2$. It is not clear if the maxima over the two subsets are asymptotically independent and hence it is not clear if we would continue to obtain the same limit. Observe that for example, if k is odd and k' is even, then

$$\mathrm{Cov}(\lambda_{k,x}, \lambda_{k',x}) = \frac{-D_{k,k'}}{n} \cot \frac{\pi(k+k')}{2n} - \frac{E_{k,k'}}{n} \cot \frac{\pi(k'-k)}{2n},$$

where $D_{k,k'}$ and $E_{k,k'}$ are as defined in (7.17). So for these covariance terms to tend to zero, we have to truncate the index set from below appropriately. For instance, if the inputs are normal, we may consider the set $L' = \{(k, k') : 1 < k < \lfloor n p_n/2 \rfloor, \; k + \lfloor n q_n/2 \rfloor < k' < \lfloor n p_n/2 \rfloor\}$ with $q_n \to 0$, and approximate it by the i.i.d. counterparts since $\sup_{k,k' \in L'} |\mathrm{Cov}(\lambda_{k,x}, \lambda_{k',x})| \to 0$ as $n \to \infty$. The complication comes when dealing with the complement of L' since it no longer has small cardinality.

7.4 k-circulant

First recall the eigenvalues of the k-circulant matrix $A_{k,n}$ from Section 1.4 of Chapter 1. For any positive integers k and n, let $p_1 < p_2 < \cdots < p_c$ be all their common prime factors so that

$$n = n' \prod_{q=1}^{c} p_q^{\beta_q} \quad \text{and} \quad k = k' \prod_{q=1}^{c} p_q^{\alpha_q}.$$

Here α_q, $\beta_q \geq 1$ and n', k', p_q are pairwise relatively prime. Then the characteristic polynomial of $A_{k,n}$ is given by

$$\chi(A_{k,n}) = \lambda^{n-n'} \prod_{j=0}^{\ell-1} (\lambda^{n_j} - y_j), \tag{7.30}$$

where y_j, n_j are as defined in Section 1.4.

7.4.1 *k*-circulant for $n = k^2 + 1$

We first consider the k-circulant matrix with $n = k^2 + 1$. In this case, clearly $n' = n$ and $k' = k$. From Lemma 4.5.2(a), $g_1 = 4$ and the eigenvalue partition of $\{0, 1, 2, \ldots, n-1\}$ contains exactly $q = \lfloor \frac{n}{4} \rfloor$ sets of size 4, say $\{\mathcal{P}_1, \mathcal{P}_2, \ldots, \mathcal{P}_{\lfloor \frac{n}{4} \rfloor}\}$. Since each $\mathcal{P}_i$ is self-conjugate, we can find a set $\mathcal{A}_i \subset \mathcal{P}_i$ of size 2 such that

$$\mathcal{P}_j = \{x : x \in \mathcal{A}_j \text{ or } n - x \in \mathcal{A}_j\}.$$

Since we shall be using the bounds from Lemma 7.2.2, we define a few relevant notations for convenience. Define

$$I_{x,n}(\omega_j) = \frac{1}{n} \left| \sum_{l=1}^{n} x_l e^{i\omega_j l} \right|^2, \quad I_{\varepsilon,n}(\omega_j) = \frac{1}{n} \left| \sum_{l=1}^{n} \varepsilon_l e^{i\omega_j l} \right|^2,$$

$$J_{x,n}(\omega) = \frac{1}{\sqrt{n}} \sum_{l=1}^{n} x_l e^{i\omega_j l}, \quad J_{\varepsilon,n}(\omega) = \frac{1}{\sqrt{n}} \sum_{l=1}^{n} \varepsilon_l e^{i\omega_j l},$$

$$\beta_{x,n}(t) = \prod_{j \in \mathcal{A}_t} I_{x,n}(\omega_j), \quad \beta_{\varepsilon,n}(t) = \prod_{j \in \mathcal{A}_t} I_{\varepsilon,n}(\omega_j),$$

$$A(\omega_j) = \sum_{t=-\infty}^{\infty} a_t e^{i\omega_j t}, \quad T_n(\omega_j) = I_{x,n}(\omega_j) - |A(\omega_j)|^2 I_{\varepsilon,n}(\omega_j),$$

$$\tilde{\beta}_{x,n}(t) = \frac{\beta_{x,n}(t)}{\prod_{j \in \mathcal{A}_t} 2\pi f(\omega_j)}, \quad \text{and} \quad M(\frac{1}{\sqrt{n}} A_{k,n}, f) = \max_{1 \leq t \leq q} \left(\tilde{\beta}_{x,n}(t) \right)^{1/4}.$$

Theorem 7.4.1 (Bose et al. (2011c)). Let $\{x_n\}$ be the two-sided moving average process defined in (7.1), and which satisfies Assumption 7.1.1. Then for $n = k^2 + 1$, as $n \to \infty$,

$$\frac{M(\frac{1}{\sqrt{n}} A_{k,n}, f) - d_q}{c_q} \xrightarrow{\mathcal{D}} \Lambda,$$

where $q = q(n) = \lfloor \frac{n}{4} \rfloor$ and c_q, d_q are same as in Theorem 6.4.1 with $g = 2$.

Proof. Observe that

$$\tilde{\beta}_{x,n}(t) := \frac{\beta_{x,n}(t)}{\prod_{j \in \mathcal{A}_t} 2\pi f(\omega_j)} = \beta_{\varepsilon,n}(t) + R_n(t),$$

where

$$R_n(t) = I_{\varepsilon,n}(\omega_{t_1})\frac{T_n(\omega_{t_2})}{2\pi f(\omega_{t_2})} + I_{\varepsilon,n}(\omega_{t_2})\frac{T_n(\omega_{t_1})}{2\pi f(\omega_{t_1})} + \frac{T_n(\omega_{t_1})}{2\pi f(\omega_{t_1})}\frac{T_n(\omega_{t_2})}{2\pi f(\omega_{t_2})}.$$

Let $q = \lfloor \frac{n}{4} \rfloor$. Recall that

$$M(\frac{1}{\sqrt{n}}A_{k,n}, f) = \max_{1 \le t \le q} (\tilde{\beta}_{x,n}(t))^{1/4}. \qquad (7.31)$$

We shall show that $\max_{1 \le t \le q} |\tilde{\beta}_{x,n}(t) - \beta_{\varepsilon,n}(t)| \to 0$ in probability. Now

$$|\tilde{\beta}_{x,n}(t) - \beta_{\varepsilon,n}(t)|$$
$$\le |I_{\varepsilon,n}(\omega_{t_1})\frac{T_n(\omega_{t_2})}{2\pi f(\omega_{t_2})}| + |I_{\varepsilon,n}(\omega_{t_2})\frac{T_n(\omega_{t_1})}{2\pi f(\omega_{t_1})}| + |\frac{T_n(\omega_{t_1})}{2\pi f(\omega_{t_1})}\frac{T_n(\omega_{t_2})}{2\pi f(\omega_{t_2})}|.$$

Note that

$$\max_{1 \le t \le q} |I_{\varepsilon,n}(\omega_{t_1})\frac{T_n(\omega_{t_2})}{2\pi f(\omega_{t_2})}| \le \frac{1}{2\pi\alpha} \max_{1 \le t < \frac{n}{2}} |I_{\varepsilon,n}(\omega_t)| \max_{1 \le t < \frac{n}{2}} |T_n(\omega_t)|.$$

From (7.7) we get

$$\max_{1 \le t \le n} |T_n(\omega_t)| = O_p(n^{-1/4}(\ln n)^{1/2}).$$

Therefore

$$\max_{1 \le t \le q} |I_{\varepsilon,n}(\omega_{t_1})\frac{T_n(\omega_{t_2})}{2\pi f(\omega_{t_2})}| = O_p(n^{-1/4}(\ln n)^{3/2}),$$

and

$$\max_{1 \le t \le q} |\frac{T_n(\omega_{t_1})}{2\pi f(\omega_{t_1})}\frac{T_n(\omega_{t_2})}{2\pi f(\omega_{t_2})}| = O_p(n^{-1/2}\ln n).$$

Combining all these we have

$$\max_{1 \le t \le q} |R_n(t)| = \max_{1 \le t \le q} |\tilde{\beta}_{x,n}(t) - \beta_{\varepsilon,n}(t)| = O_p(n^{-1/4}(\ln n)^{3/2}).$$

Note that

$$(\beta_{\varepsilon,n}(t))^{1/4} - |R_n(t)|^{1/4} \le (\tilde{\beta}_{x,n}(t))^{1/4} \le (\beta_{\varepsilon,n}(t))^{1/4} + |R_n(t)|^{1/4},$$

and hence

$$|\max_{1 \le t \le q} (\tilde{\beta}_{x,n}(t))^{1/4} - \max_{1 \le t \le q} (\beta_{\varepsilon,n}(t))^{1/4}| = O_p(n^{-1/16}(\ln n)^{3/8}). \qquad (7.32)$$

From Theorem 6.4.1, we know that

$$\frac{\max_{1 \leq t \leq q} \left(\beta_{\varepsilon,n}(t)\right)^{1/4} - d_q}{c_q} \xrightarrow{\mathcal{D}} \Lambda. \tag{7.33}$$

Hence, from (7.31), (7.32) and (7.33) it follows that

$$\frac{M(\frac{1}{\sqrt{n}} A_{k,n}, f) - d_q}{c_q} \xrightarrow{\mathcal{D}} \Lambda.$$

$\square$

7.4.2   *k*-circulant for $n = k^g + 1$, $g > 2$

Now we extend Theorem 7.4.1 for $n = k^g + 1$ where $g > 2$. Here, we use a slightly different notation to use the developments of Section 6.2 of Chapter 6. Define

$$\tilde{\beta}_{x,j}(t) := \frac{\beta_{x,j}(t)}{\prod_{l \in \mathcal{A}_t} 2\pi f(\omega_l)} \quad \text{and} \quad M(\frac{1}{\sqrt{n}} A_{k,n}, f) = \max_l \max_{j:\mathcal{P}_j \in J_l} \left(\tilde{\beta}_{x,l}(j)\right)^{1/2l}.$$

Theorem 7.4.2 (Bose et al. (2011c)). *Let* $\{x_n\}$ *be the two-sided moving average process defined in (7.1) and which satisfies Assumption 7.1.1. Then for* $n = k^g + 1$, $g > 2$, *as* $n \to \infty$,

$$\frac{M(\frac{1}{\sqrt{n}} A_{k,n}, f) - d_q}{c_q} \xrightarrow{\mathcal{D}} \Lambda,$$

where $q = q(n) = \frac{n}{2g}$, *and* c_q, d_q *are as defined in Theorem 6.4.1.*

Proof. The line of argument is similar to that in the case of $g = 2$. To prove the result we use following two facts:

(i) From (7.7),

$$\max_{1 \leq t < \frac{n}{2}} |T_n(\omega_t)| = o_p(n^{-1/4}(\ln n)^{1/2}).$$

(ii) From Davis and Mikosch (1999) (see Theorem 11.3.2 in Appendix),

$$\max_{1 \leq t < \frac{n}{2}} |I_{\varepsilon,n}(\omega_t)| = O_p(\ln n) \quad \text{and} \quad \max_{1 \leq t < \frac{n}{2}} |I_{x,n}(\omega_t)| = O_p(\ln n).$$

Using these, and inequality (6.28) of Chapter 6, it is easy to see that, for some $\delta_0 > 0$,

$$\max_l \max_{j:\mathcal{P}_j \in J_l} \left|\tilde{\beta}_{x,l}(t) - \beta_{\varepsilon,l}(t)\right| = o_p(n^{-\delta_0}). \tag{7.34}$$

Now the results follow from Theorem 6.4.1 and (7.34). $\square$

7.5   Exercises

1. Prove the results mentioned in Remark 7.2.1.

2. Complete the proof of (7.10).

3. Prove (7.19) of Lemma 7.3.5.

4. Prove (7.24).

5. Prove (7.28) in the proof of Theorem 7.3.3.

6. Show that under the conditions of Theorem 7.3.6,

$$\frac{\max_{1 \leq k \leq \lfloor \frac{n}{2} \rfloor} \frac{|\lambda_{k,x}|}{\sqrt{2\pi f(\omega_k)}}}{\sqrt{\ln n}} \xrightarrow{P} 1.$$

   Hint: The proof is similar to the proof of Theorem 7.3.6, with the normalizing constants changed suitably.

7. If we include λ_0 in the definition of $M(\frac{1}{\sqrt{n}}SC_n, f)$, that is, if $\overline{M}(\frac{1}{\sqrt{n}}SC_n, f) = \max_{0 \leq k \leq \lfloor \frac{n}{2} \rfloor} \frac{|\lambda_k|}{\sqrt{2\pi f(\omega_k)}}$, then show that under Assumption 7.1.1 except that mean μ of $\{\varepsilon_i\}$ is now non-zero,

$$\overline{M}(\frac{1}{\sqrt{n}}SC_n, f) - |\mu|\sqrt{n} \xrightarrow{D} N(0, 2).$$

8

Poisson convergence

So far we have studied the asymptotic behavior of circulant-type random matrices through the bulk distribution of the eigenvalues (LSD) and the distribution of the extremes (spectral radius). It seems then natural to study the joint behavior of the eigenvalues via the point process approach.

The most appropriate point process for circulant type matrices is the one based on the points $\{(\omega_k, \frac{\lambda_k - b_q}{a_q}), 0 \leq k < n\}$ where $\{\lambda_k\}$ are the appropriately labelled eigenvalues, $\{\omega_k = \frac{2\pi k}{n}\}$ are the Fourier frequencies, and a_q, b_q are appropriate scaling and centering constants that appeared in the weak convergence of the spectral radius in Chapter 5. In this chapter we study their asymptotic behavior in the case of i.i.d. light-tailed entries.

In each case the limit measure turns out to be Poisson. In particular, this yields the distributional convergence of any k-upper ordered eigenvalues of these matrices, and also yields the joint distributional convergence of any fixed number of spacings of the upper ordered eigenvalues.

8.1 Point process

Let E be any set equipped with an appropriate sigma-algebra $\mathcal{E}$. For our purposes, usually $E = ([0, \pi] \times (-\infty, \infty])$ and is equipped with the Borel sigma-algebra. Let $M(E)$ be the space of all *point measures* on E, endowed with the topology of *vague convergence*. Any *point process* on E is a measurable map

$$N : (\Omega, \mathcal{F}, \mathrm{P}) \to (M(E), \mathcal{M}(E)).$$

It is said to be *simple* if

$$\mathrm{P}(N(\{x\}) \leq 1, x \in E) = 1.$$

Let $\xrightarrow{\mathcal{V}}$ denote the convergence in distribution relative to the vague topology (see Section 11.2 of Appendix). The following result provides a criterion for convergence of point processes. Its proof is available in Kallenberg (1986), Resnick (1987) and Embrechts et al. (1997). This lemma will play a key role in the proofs of our results.

Lemma 8.1.1. Let $\{N_n\}$ be a sequence of point processes and N be a simple point process on a complete separable metric space E. Let $\mathcal{T}$ be a basis of relatively compact open sets such that $\mathcal{T}$ is closed under finite unions and intersections, and for $I \in \mathcal{T}$, $P[N(\partial I) = 0] = 1$. If $\lim_{n\to\infty} P[N_n(I) = 0] = P[N(I) = 0]$ and $\lim_{n\to\infty} E[N_n(I)] = E[N(I)] < \infty$, then $N_n \xrightarrow{\mathcal{V}} N$ in $M(E)$.

Very few studies exist on the point process of eigenvalues of general random matrices. As an example, Soshnikov (2004) considered the point process based on the positive eigenvalues of an appropriately scaled Wigner matrix with entries $\{x_{ij}\}$ which satisfy $P(|x_{ij}| > x) = h(x)x^{-\alpha}$, where h is a slowly varying function at infinity and $0 < \alpha < 2$. He showed that it converges to an inhomogeneous Poisson random point process. A similar result was proved for sample covariance matrices with Cauchy entries in Soshnikov (2006). These results on the Wigner and the sample covariance matrices were extended in Auffinger et al. (2009) for $2 \leq \alpha < 4$.

8.2   Reverse circulant

Earlier we used the labelling λ_k, $\lambda_{n,k}$ etc. for the eigenvalues. Now it will be convenient to re-label them as $\lambda_{n,x}(\omega_k)$ for the input sequence $\{x_i\}$. Hence the eigenvalues of $\frac{1}{\sqrt{n}}RC_n$ with the input sequence $\{x_i\}$ (see Section 1.3 of Chapter 1) are now written as:

$$
\begin{cases}
\lambda_{n,x}(\omega_0) & = \dfrac{1}{\sqrt{n}} \displaystyle\sum_{t=0}^{n-1} x_t \\[2mm]
\lambda_{n,x}(\omega_{n/2}) & = \dfrac{1}{\sqrt{n}} \displaystyle\sum_{t=0}^{n-1} (-1)^t x_t, \text{ if } n \text{ is even} \\[2mm]
\lambda_{n,x}(\omega_k) & = -\lambda_{n,x}(\omega_{n-k}) = \sqrt{I_{n,x}(\omega_k)}, \ 1 \leq k \leq \lfloor \dfrac{n-1}{2} \rfloor,
\end{cases}
\tag{8.1}
$$

where

$$
I_{n,x}(\omega_k) = \frac{1}{n} \Big| \sum_{t=0}^{n-1} x_t e^{-it\omega_k} \Big|^2 \quad \text{and} \quad \omega_k = \frac{2\pi k}{n}.
$$

Since the eigenvalues occur in pairs with opposite signs (except when n is odd), it suffices to define our point process based on the points $(\omega_k, \frac{\lambda_{n,x}(\omega_k) - b_q}{a_q})$ for $k = 0, 1, 2, \ldots, \lfloor \frac{n}{2} \rfloor$. Let $\epsilon_a(\cdot)$ denote the point measure which gives unit mass to any set containing a. Motivated by the scaling and centering constants in Chapter 5, we define

$$
\eta_n(\cdot) = \sum_{j=0}^{q} \epsilon_{\left(\omega_j, \frac{\lambda_{n,x}(\omega_j) - b_q}{a_q}\right)}(\cdot),
\tag{8.2}
$$

where $q = q(n) = \lfloor \frac{n}{2} \rfloor$, $a_q = \frac{1}{2\sqrt{\ln q}}$ and $b_q = \sqrt{\ln q}$. Throughout this chapter, the input sequence is assumed to satisfy the following assumption.

Assumption 8.2.1. $\{x_i\}$ are i.i.d., $E[x_0] = 0$, $E[x_0]^2 = 1$, and $E|x_0|^s < \infty$ for some $s > 2$.

Theorem 8.2.1 (Bose et al. (2011b)). Suppose $\{x_i\}$ satisfies Assumption 8.2.1. Let η_n be as in (8.2) and let η be a Poisson point process on $[0, \pi] \times (-\infty, \infty]$ with intensity function $\lambda(t, x) = \pi^{-1}e^{-x}$. Then $\eta_n \xrightarrow{\mathcal{V}} \eta$.

The main ideas of the proof are germane in the normal approximation methods that we have already encountered. When the input sequence is i.i.d. normal, the positive eigenvalues of RC_n are independent and each is distributed as the square root of exponential (see Lemma 3.3.2), and the convergence follows immediately.

When the entries are not normal, the sophisticated normal approximation given in Lemma 6.3.2 allows us to replace the variables by appropriate normal variables after requisite truncation. Let

$$\bar{x}_t = x_t \mathbb{I}(|x_t| < n^{1/s}) - E[x_t \mathbb{I}(|x_t| < n^{1/s})].$$

Let $\{N_t\}$ be a sequence of i.i.d. $N(0, 1)$ random variables, and ϕ_{C_d} be the density of the d-dimensional centered normal random vector with covariance matrix C_d. For $d \geq 1$, and distinct Fourier frequencies $\omega_{i_1}, \ldots, \omega_{i_d}$, let

$$v_d(t) = (\cos(\omega_{i_1}t), \sin(\omega_{i_2}t), \ldots, \cos(\omega_{i_d}t), \sin(\omega_{i_d}t))'. \qquad (8.3)$$

Now consider RC_n with the input sequences $\{\bar{x}_t + \sigma_n N_t\}$ and $\{\bar{x}_t\}$, where σ_n will be chosen later appropriately. Let η_n^* and $\bar{\eta}_n$ be the respective point processes. For technical convenience, while defining η_n^*, we consider the distinct eigenvalues and leave out λ_0. Then we proceed to show that as $n \to \infty$,

(i) η_n^* converges to η, and

(ii) η_n^* and η_n are close in probability via $\bar{\eta}_n$.

This is essentially the program that is carried out for other matrices also. Finally, the dependent case is reduced to the independent case by an appropriate approximation result such as Lemma 7.2.2.

Proof of Theorem 8.2.1. The proof is done in two steps.

Step 1: We first show that $\eta_n^* \xrightarrow{\mathcal{V}} \eta$ where

$$\eta_n^*(\cdot) = \sum_{j=1}^{q} \epsilon_{\left(\omega_j, \frac{\lambda_{n,\bar{x}+\sigma_n N}(\omega_j) - b_q}{a_q}\right)}(\cdot),$$

and $\lambda_{n,\bar{x}+\sigma_n N}(\omega_k)$ are the eigenvalues of $\frac{1}{\sqrt{n}} RC_n$ with entries $\{\bar{x}_t + \sigma_n N_t\}$

where $\sigma_n^2 = n^{-c}$ and c is as in Lemma 6.3.2. First note that if we define the set

$$A_q^d = \{(x_1, y_1, \ldots, x_d, y_d)' : \sqrt{x_i^2 + y_i^2} > 2z_q\},$$

where $z_q = a_q x + b_q$, it easily follows that

$$
\begin{aligned}
& \mathrm{P}\left(\lambda_{n,\bar{x}+\sigma_n N}(\omega_{i_1}) > z_q, \ldots, \lambda_{n,\bar{x}+\sigma_n N}(\omega_{i_d}) > z_q\right) \\
&= \quad \mathrm{P}\left(2^{1/2} \frac{1}{\sqrt{n}} \sum_{t=1}^{n} (\bar{x}_t + \sigma_n N_t) v_d(t) \in A_q^d\right) \\
&= \quad \int_{A_q^d} \phi_{(1+\sigma_n^2) I_{2d}}(x)(1 + o(1)) dx \\
&= \quad q^{-d} \exp(-dx)(1 + o(1)).
\end{aligned}
\tag{8.4}
$$

Since the limit process η is simple, by Lemma 8.1.1, it suffices to show that

$$\mathrm{E}\,\eta_n^*((a, b] \times (x, y]) \to \mathrm{E}\,\eta((a, b] \times (x, y]) = \frac{b-a}{\pi}(e^{-x} - e^{-y}), \tag{8.5}$$

for all $0 \le a < b \le \pi$ and $x < y$, and for all $k \ge 1$,

$$
\begin{aligned}
& \mathrm{P}(\eta_n^*((a_1, b_1] \times R_1) = 0, \ldots, \eta_n^*((a_k, b_k] \times R_k) = 0) \\
& \to \mathrm{P}(\eta((a_1, b_1] \times R_1) = 0, \ldots, \eta((a_k, b_k] \times R_k) = 0),
\end{aligned}
\tag{8.6}
$$

where $0 \le a_1 < b_1 < \cdots < a_k < b_k \le \pi$, and $R_1, \ldots, R_k$ are bounded Borel sets, each of which consists of a finite union of intervals of $(-\infty, \infty]$.

Proof of (8.5): This relation is established as follows:

$$
\begin{aligned}
\mathrm{E}\,\eta_n^*((a, b] \times (x, y]) &= \sum_{\omega_j \in (a, b]} \mathrm{P}(a_q x + b_q < \lambda_{n,\bar{x}+\sigma_n N}(\omega_j) \le a_q y + b_q) \\
\text{(by (8.4))} \quad &\sim \frac{(b-a)n}{2\pi} q^{-1}(e^{-x} - e^{-y}) \to \frac{(b-a)}{\pi}(e^{-x} - e^{-y}).
\end{aligned}
$$

Proof of (8.6): Set $n_j := \#\{i : \omega_i \in (a_j, b_j]\} \sim n(b_j - a_j)$. The complement of the event in (8.6) is the union of $m = n_1 + \cdots + n_k$ events, so that

$$
\begin{aligned}
& 1 - \mathrm{P}(\eta_n^*((a_1, b_1] \times R_1) = 0, \ldots, \eta_n^*((a_k, b_k] \times R_k) = 0) \\
&= \quad \mathrm{P}\Big(\bigcup_{j=1}^{k} \bigcup_{\omega_i \in (a_j, b_j]} \{\frac{\lambda_{n,\bar{x}+\sigma_n N}(\omega_i) - b_q}{a_q} \in R_j\}\Big).
\end{aligned}
\tag{8.7}
$$

Now for any choice of d distinct integers $i_1, \ldots, i_d \in \{1, \ldots, q\}$ and integers $j_1, \ldots, j_d \in \{1, \ldots, k\}$, we have from (8.4) that

$$\mathrm{P}\Big(\bigcap_{r=1}^{d} \{\frac{\lambda_{n,\bar{x}+\sigma_n N}(\omega_{i_r}) - b_q}{a_q} \in R_{j_r}\}\Big) = q^{-d} \prod_{r=1}^{d} \Lambda(R_{j_r})(1 + o(1)), \tag{8.8}$$

where $\Lambda(B) = \int_B e^{-x} dx$, B bounded measurable, and (8.8) holds uniformly over all d-tuples $i_1, \ldots, i_d$. Using this and an elementary counting argument, the sum of the probabilities of all collections of d distinct sets from the m sets that comprise the union in (8.7) equals,

$$S_d$$

$$= \sum_{\substack{(u_1, \ldots, u_k), \\ u_1 + \cdots + u_k = d}} \binom{n_1}{u_1} \cdots \binom{n_k}{u_k} q^{-u_1} \lambda^{u_1}(R_1) \cdots q^{-u_k} \lambda^{u_k}(R_k)(1 + o(1))$$

$$= \sum_{\substack{(u_1, \ldots, u_k), \\ u_1 + \cdots + u_k = d}} \frac{1}{u_1! \cdots u_k! \pi^d} ((b_1 - a_1)\lambda(R_1))^{u_1} \cdots ((b_k - a_k)\lambda(R_k))^{u_k}(1 + o(1))$$

$$\to (d!)^{-1} \pi^{-d} ((b_1 - a_1)\lambda(R_1) + \cdots + (b_k - a_k)\lambda(R_k))^d.$$

Now it follows that

$$\sum_{j=1}^{2s} (-1)^{j-1} S_j \overset{n \to \infty}{\Longrightarrow} \sum_{j=1}^{2s} \frac{(-1)^{j-1}}{j! \pi^j} ((b_1 - a_1)\lambda(R_1) + \ldots + (b_k - a_k)\lambda(R_k))^j$$

$$\overset{s \to \infty}{\Longrightarrow} 1 - \exp\Big(-\sum_{j=1}^{k}(b_j - a_j)\pi^{-1}\lambda(R_j)\Big),$$

which by Bonferroni inequality and (8.7), proves (8.6).

Step 2: It remains to show that η_n^* is close to η_n. Define

$$\bar{\eta}_n(\cdot) = \sum_{j=1}^{q} \epsilon_{\left(\omega_j, \frac{\lambda_{n,\bar{x}}(\omega_j) - b_q}{a_q}\right)}(\cdot) \quad \text{and} \quad \eta_n'(\cdot) = \sum_{j=1}^{q} \epsilon_{\left(\omega_j, \frac{\lambda_{n,x}(\omega_j) - b_q}{a_q}\right)}(\cdot).$$

It then suffices to show that (see Theorem 4.2 of Kallenberg (1986))

$$\bar{\eta}_n - \eta_n^* \overset{P}{\longrightarrow} 0, \tag{8.9}$$

$$\bar{\eta}_n - \eta_n' \overset{P}{\longrightarrow} 0, \text{ and} \tag{8.10}$$

$$\eta_n' - \eta_n \overset{P}{\longrightarrow} 0. \tag{8.11}$$

For this, it is enough to show that for any continuous function f on $[0, \pi] \times (-\infty, \infty]$ with compact support,

$$\bar{\eta}_n(f) - \eta_n^*(f) \overset{P}{\longrightarrow} 0, \ \bar{\eta}_n(f) - \eta_n'(f) \overset{P}{\longrightarrow} 0, \text{ and } \eta_n'(f) - \eta_n(f) \overset{P}{\longrightarrow} 0,$$

where $\eta(f)$ denotes $\int f d\eta$. Suppose that

$$\text{support of } f \subseteq [0, \pi] \times [K + \gamma_0, \infty), \text{ for some } \gamma_0 > 0 \text{ and } K \in \mathbb{R}. \tag{8.12}$$

Define

$$\omega(\gamma) := \sup\{|f(t, x) - f(t, y)|; \ t \in [0, \pi], \ |x - y| \leq \gamma\}.$$

Note that, since f is uniformly continuous,

$$\omega(\gamma) \to 0 \ \text{ as } \gamma \to 0.$$

Proof of (8.9): Let

$$A_n = \{ \max_{j=1,\dots,q} |\frac{\lambda_{n,\bar{x}+\sigma_n N}(\omega_j)}{a_q} - \frac{\lambda_{n,\bar{x}}(\omega_j)}{a_q}| \le \gamma \}.$$

Then we have for $\gamma < \gamma_0$,

$$|f(\omega_j, \frac{\lambda_{n,\bar{x}+\sigma_n N}(\omega_j) - b_q}{a_q}) - f(\omega_j, \frac{\lambda_{n,\bar{x}}(\omega_j) - b_q}{a_q})|$$

$$\le \begin{cases} \omega(\gamma) & \text{if } \frac{\lambda_{n,\bar{x}+\sigma_n N}(\omega_j) - b_q}{a_q} > K \\ 0 & \text{if } \frac{\lambda_{n,\bar{x}+\sigma_n N}(\omega_j) - b_q}{a_q} \le K. \end{cases} \tag{8.13}$$

Also note that

$$\frac{1}{a_q} \max_{1 \le j \le q} |\lambda_{n,\bar{x}+\sigma_n N}(\omega_j) - \lambda_{n,\bar{x}}(\omega_j)|$$

$$\le \frac{1}{a_q} \max_{1 \le j \le q} |\frac{\sigma_n}{\sqrt{n}} \sum_{t=1}^{n} N_t e^{i\omega_j t}|$$

$$\le \frac{\sigma_n}{a_q} \max_{1 \le j \le q} \sqrt{\frac{1}{n}(\sum_{t=1}^{n} N_t \cos \frac{2\pi k t}{n})^2 + \frac{1}{n}(\sum_{t=1}^{n} N_t \sin \frac{2\pi k t}{n})^2}$$

$$\le \frac{\sigma_n}{a_q} \max_{1 \le j \le q} \sqrt{X_{1j}^2 + X_{2j}^2},$$

where $\{X_{1j}, X_{2j}; 1 \le j \le q\}$ are i.i.d. $N(0,1)$. Now,

$$\frac{\sigma_n}{a_q} \max_{1 \le j \le q} \sqrt{X_{1j}^2 + X_{2j}^2} = O_P(\sigma_n \ln n).$$

Therefore $\lim_{n \to \infty} P(A_n^c) = 0$. For any $\epsilon > 0$, choose $\gamma > 0$ so that $\gamma < \gamma_0$. Define

$$B_n = \{|\bar{\eta}_n(f) - \eta_n^*(f)| > \epsilon\}.$$

Then

$$\begin{aligned} \limsup_{n \to \infty} P(B_n) &\le \limsup_{n \to \infty}(P(B_n \cap A_n) + P(A_n^c)) \\ &\le \limsup_{n \to \infty} P(\omega(\gamma)\eta_n^*([0,\pi] \times [K,\infty)) > \epsilon) + \limsup_{n \to \infty} P(A_n^c) \\ &\le \limsup_{n \to \infty} E\,\eta_n^*([0,\pi] \times [K,\infty))\omega(\gamma)/\epsilon \\ &\le e^{-K}\omega(\gamma)/\epsilon \to 0 \ \text{ as } \gamma \to 0. \end{aligned}$$

This completes the proof of (8.9).

The proof of (8.10) is essentially as above and we omit it.

Proof of (8.11): Finally for any $\epsilon > 0$,

$$
\begin{aligned}
P(|\eta_n'(f) - \eta_n(f)| > \epsilon) &= P(|f(0, \frac{\lambda_{n,x}(\omega_0) - b_q}{a_q})| > \epsilon) \\
&\leq P(\frac{\lambda_{n,x}(\omega_0) - b_q}{a_q} \geq K) \\
&= P(\frac{1}{\sqrt{n}} \sum_{l=0}^{n-1} x_l > K a_q + b_q) \to 0, \text{ as } n \to \infty.
\end{aligned}
$$

The proof of Theorem 8.2.1 is now complete. □

Let $\lambda_{n,(q)} \leq \cdots \leq \lambda_{n,(2)} \leq \lambda_{n,(1)}$ be the ordered eigenvalues. Then for any fixed k, the joint limit distribution of k-upper ordered eigenvalues as well as their spacings can be derived from Theorem 8.2.1.

Corollary 8.2.2. Under the assumption of Theorem 8.2.1,

(a) for any real numbers $x_k < \cdots < x_2 < x_1$,

$$
P\left(\frac{\lambda_{n,(1)} - b_q}{a_q} \leq x_1, \ldots, \frac{\lambda_{n,(k)} - b_q}{a_q} \leq x_k\right) \to P(Y_{(1)} \leq x_1, \ldots, Y_{(k)} \leq x_k),
$$

where $(Y_{(1)}, \ldots, Y_{(k)})$ has the density $\exp(-\exp(-x_k) - (x_1 + \cdots + x_{k-1}))$.

(b) $\left(\frac{\lambda_{n,(i)} - \lambda_{n,(i-1)}}{a_q}\right)_{i=1,\ldots,k} \xrightarrow{D} (i^{-1} E_i)_{i=1,\ldots,k}$, where $\{E_i\}$ is a sequence of i.i.d. standard exponential random variables.

Proof. The proof is similar to the proof of Theorem 4.2.8 of Embrechts et al. (1997). We just briefly sketch the steps. Let $x_k < \cdots < x_1$ be real numbers, and write $N_{i,n} = \eta_n([0, \pi] \times (x_i, \infty))$ for the number of *exceedances* of x_i by $\frac{\lambda_{n,x}(\omega_j) - b_q}{a_q}$, $j = 1, \ldots, q$. Then

$$
\begin{aligned}
P\left(\frac{\lambda_{n,(1)} - b_q}{a_q} \leq x_1, \ldots, \frac{\lambda_{n,(k)} - b_q}{a_q} \leq x_k\right) \\
= P(N_{1,n} = 0, N_{2,n} \leq 1, \ldots, N_{k,n} \leq k - 1) \\
\to P(N_1 = 0, N_2 \leq 1, \ldots, N_k \leq k - 1),
\end{aligned}
$$

where $N_i = \eta([0, \pi] \times (x_i, \infty])$. Let $Z_i = \eta([0, \pi] \times (x_i, x_{i-1}])$ with $x_0 = \infty$. To calculate the above probability, it is enough to consider $P(N_1 = a_1, N_2 = a_1 + a_2, \ldots, N_k = a_1 + \cdots + a_k)$, where $a_i \geq 0$. However,

$$
\begin{aligned}
&P(N_1 = a_1, N_2 = a_1 + a_2, \ldots, N_k = a_1 + \cdots + a_k) \\
&= P(Z_1 = a_1, Z_2 = a_2, \ldots, Z_k = a_k) \\
&= \frac{(e^{-x_1})^{a_1}}{a_1!} \frac{(e^{-x_2} - e^{-x_1})^{a_2}}{a_2!} \cdots \frac{(e^{-x_k} - e^{-x_{k-1}})^{a_k}}{a_k!} e^{-e^{-x_k}}.
\end{aligned}
$$

This proves Part (a). Part (b) is an easy consequence of Part (a).   □

8.3   Symmetric circulant

The eigenvalues of $\frac{1}{\sqrt{n}}SC_n$ (we now index them by $\{\omega_k\}$) are given by (see Section 1.2 of Chapter 1):
(i) for n odd:

$$
\begin{cases}
\lambda_{n,x}(\omega_0) &= \dfrac{1}{\sqrt{n}}\Big[x_0 + 2\displaystyle\sum_{j=1}^{\lfloor \frac{n}{2} \rfloor} x_j\Big] \\[4mm]
\lambda_{n,x}(\omega_k) &= \dfrac{1}{\sqrt{n}}\Big[x_0 + 2\displaystyle\sum_{j=1}^{\lfloor \frac{n}{2} \rfloor} x_j \cos(\omega_k j)\Big], \quad 1 \le k \le \Big\lfloor \dfrac{n}{2} \Big\rfloor
\end{cases}
\tag{8.14}
$$

(ii) for n even:

$$
\begin{cases}
\lambda_{n,x}(\omega_0) &= \frac{1}{\sqrt{n}}\big[x_0 + 2\sum_{j=1}^{\frac{n}{2}-1} x_j + x_{n/2}\big] \\[3mm]
\lambda_{n,x}(\omega_k) &= \frac{1}{\sqrt{n}}\big[x_0 + 2\sum_{j=1}^{\frac{n}{2}-1} x_j \cos(\omega_k j) + (-1)^k x_{n/2}\big], \quad 1 \le k \le \frac{n}{2},
\end{cases}
$$

with $\lambda_{n,x}(\omega_{n-k}) = \lambda_{n,x}(\omega_k)$ in both cases.

Now define a sequence of point processes based on the points $(\omega_j, \frac{\lambda_{n,x}(\omega_j) - b_q}{a_q})$ for $k = 0, 1, \ldots, q (= \lfloor \frac{n}{2} \rfloor)$, where $\lambda_{n,x}(\cdot)$ are as in (8.14). Note that we are not considering the eigenvalues λ_{n-k} for $k = 1, \ldots, \lfloor \frac{n}{2} \rfloor$ since $\lambda_{n,x}(\omega_{n-k}) = \lambda_{n,x}(\omega_k)$ for $k = 1, \ldots, \lfloor \frac{n}{2} \rfloor$. Define

$$
\eta_n(\cdot) = \sum_{j=0}^{q} \epsilon_{\left(\omega_j, \frac{\lambda_{n,x}(\omega_j) - b_q}{a_q}\right)}(\cdot),
\tag{8.15}
$$

where

$$
b_n = c_n + a_n \ln 2, \quad a_n = (2\ln n)^{-1/2} \text{ and } c_n = (2\ln n)^{1/2} - \frac{\ln\ln n + \ln 4\pi}{2(2\ln n)^{1/2}}.
\tag{8.16}
$$

Theorem 8.3.1 (Bose et al. (2011b)). Let $\{x_t\}$ be i.i.d. random variables which satisfy Assumption 8.2.1. Then $\eta_n \overset{\mathcal{V}}{\to} \eta$, where η is a Poisson point process on $[0, \pi] \times (-\infty, \infty]$ with intensity function $\lambda(t, x) = \pi^{-1} e^{-x}$.

Proof of Theorem 8.3.1. The proof is similar to that of Theorem 8.2.1. So we sketch it. Let η_n^* be the point process based on $(\omega_j, \frac{\lambda'_{n,\bar{x}+\sigma_n N}(\omega_j) - b_q}{a_q})$ for $1 \leq j \leq q$, where

$$\lambda'_{n,\bar{x}+\sigma_n N}(\omega_j) = \frac{1}{\sqrt{n}}\Big[\sqrt{2}(\bar{x}_0 + \sigma_n N_0) + 2\sum_{t=1}^{\lfloor \frac{n}{2} \rfloor}(\bar{x}_t + \sigma_n N_t)\cos\frac{2\pi jt}{n}\Big], \quad 0 \leq j \leq \lfloor \frac{n}{2} \rfloor.$$

Then by verifying (8.5) and (8.6) it follows that $\eta_n^* \overset{D}{\to} \eta$.

Now define the point processes,

$$\bar{\eta}'_n(\cdot) = \sum_{j=1}^{q} \epsilon_{\big(\omega_j, \frac{\lambda'_{n,\bar{x}}(\omega_j) - b_q}{a_q}\big)}(\cdot), \quad \bar{\eta}_n(\cdot) = \sum_{j=1}^{q} \epsilon_{\big(\omega_j, \frac{\lambda_{n,\bar{x}}(\omega_j) - b_q}{a_q}\big)}(\cdot),$$

$$\eta'_n(\cdot) = \sum_{j=1}^{q} \epsilon_{\big(\omega_j, \frac{\lambda_{n,x}(\omega_j) - b_q}{a_q}\big)}(\cdot),$$

where

$$\lambda'_{n,\bar{x}}(\omega_j) = \frac{1}{\sqrt{n}}\Big[\sqrt{2}\bar{x}_0 + 2\sum_{t=1}^{\lfloor \frac{n}{2} \rfloor}\bar{x}_t\cos\frac{2\pi jt}{n}\Big], \quad 0 \leq j \leq \lfloor \frac{n}{2} \rfloor,$$

and $\{\lambda_{n,\bar{x}}(\omega_j)\}$ are as in (8.14) with x_t replaced by $\bar{x}_t$. It now suffices to show that

$$\bar{\eta}'_n - \eta_n^* \overset{P}{\to} 0, \quad \bar{\eta}_n - \bar{\eta}'_n \overset{P}{\to} 0, \quad \bar{\eta}_n - \eta'_n \overset{P}{\to} 0 \quad \text{and} \quad \eta'_n - \eta_n \overset{P}{\to} 0. \quad (8.17)$$

For the first relation in (8.17), define

$$A_n = \{\max_{1 \leq j \leq q} |\lambda'_{n,\bar{x}}(\omega_j) - \lambda_{n,\bar{x}+\sigma_n N}(\omega_j)| \leq \gamma\},$$

and observe that

$$\max_{1 \leq j \leq q} |\lambda'_{n,\bar{x}}(\omega_j) - \lambda_{n,\bar{x}+\sigma_n N}(\omega_j)| = \frac{\sigma_n}{\sqrt{n}} \max_{1 \leq j \leq q} |\sqrt{2}N_0 + 2\sum_{t=1}^{\lfloor n/2 \rfloor} N_t\cos\frac{2\pi jt}{n}|$$

$$= O_p(\sigma_n \ln n).$$

Hence $P(A_n^c) \to 0$. The remaining argument is similar to the proof of (8.9). For the second relation, note that

$$P\big(\max_{1 \leq j \leq q}|\lambda_{n,\bar{x}}(\omega_j) - \lambda'_{n,\bar{x}}(\omega_j)| > \epsilon\big) \leq P\big(\frac{(\sqrt{2}-1)|x_0|}{\sqrt{n}} > \epsilon\big) \to 0.$$

Proofs of the third and the fourth relations are similar to the proofs of (8.10) and (8.11) in the proof of Theorem 8.2.1. $\qquad\square$

The proof of the following corollary is left as an exercise.

Corollary 8.3.2. Under the assumption of Theorem 8.3.1,

(a) for any real numbers $x_k < \cdots < x_2 < x_1$,

$$P\left(\frac{\lambda_{n,(1)} - b_q}{a_q} \le x_1, \ldots, \frac{\lambda_{n,(k)} - b_q}{a_q} \le x_k\right) \to P(Y_{(1)} \le x_1, \ldots, Y_{(k)} \le x_k),$$

where $(Y_{(1)}, \ldots, Y_{(k)})$ has the density $\exp(-\exp(-x_k) - (x_1 + \cdots + x_{k-1}))$.

(b)

$$\left(\frac{\lambda_{n,(i)} - \lambda_{n,(i-1)}}{a_q}\right)_{1 \le i \le k} \xrightarrow{D} (i^{-1} E_i)_{1 \le i \le k},$$

where $\{E_i\}$ is a sequence of i.i.d. standard exponential random variables.

8.4 k-circulant, $n = k^2 + 1$

One can consider an appropriate point process based on the eigenvalues of the k-circulant matrix for $n = k^g + 1$ where $g > 2$, and can prove a result similar to Theorem 8.4.2. But for general $g > 2$, algebraic details are quite complicated. For simplicity, we consider the k-circulant matrix only for $n = k^2 + 1$.

From Lemma 4.5.2 and (4.29) of Chapter 4, $g_1 = 4$, the eigenvalue partition of $\{0, 1, 2, \ldots, n - 1\}$ contains exactly $q = \lfloor \frac{n}{4} \rfloor$ sets of size 4, and each set is self-conjugate. Moreover, if k is even then there is only one more partition set containing only 0, and if k is odd then there are two more partition sets containing only 0 and only $n/2$, respectively.

To define an appropriate point process, we need a clear picture of the eigenvalue partition of

$$\mathbb{Z}_n = \{0, 1, 2, \ldots, n - 1\}.$$

First write $S(x)$ defined in (1.5) of Chapter 1 as follows:

$$S(ak + b) = \{ak + b, bk - a, n - ak - b, n - bk + a\}; \ 0 \le a \le k - 1, \ 1 \le b \le k.$$

Lemma 8.4.1. For $n = k^2 + 1$,

$$\mathbb{Z}_n = \bigcup_{0 \le a \le \lfloor \frac{k-2}{2} \rfloor, a+1 \le b \le k-a-1} S(ak + b) \bigcup S(0), \text{ if } k \text{ is even}$$

$$= \bigcup_{0 \le a \le \lfloor \frac{k-2}{2} \rfloor, a+1 \le b \le k-a-1} S(ak + b) \bigcup S(0) \bigcup S(n/2), \text{ if } k \text{ is odd},$$

where all $S(ak + b)$ are disjoint and form an eigenvalue partition of $\mathbb{Z}_n$.

Proof. First observe that $S(0) = \{0\}$ and $S(n/2) = \{n/2\}$ if k is odd, and

$$\#\{x : x \in S(ak + b); 0 \le a \le \lfloor \frac{k-2}{2} \rfloor, a + 1 \le b \le k - a - 1\}$$

$$= \begin{cases} n - 1 & \text{if } k \text{ is even} \\ n - 2 & \text{if } k \text{ is odd.} \end{cases}$$

So if we can show that $S(ak + b); 0 \le a \le \lfloor \frac{k-2}{2} \rfloor, a + 1 \le b \le k - a - 1$ are mutually disjoint then we are done. We shall show $S(a_1 k + b_1) \cap S(a_2 k + b_2) = \emptyset$ for $a_1 \ne a_2$ or $b_1 \ne b_2$. We divide the proof into four different cases.

Case (i) $(a_1 < a_2, b_1 > b_2)$ Note that

$$a_1 + 1 < a_2 + 1 \le b_2 < b_1 \le k - (a_1 + 1).$$

Since $\{S(x); 0 \le x \le n - 1\}$ forms a partition of $\mathbb{Z}_n$, it is enough to show that $a_1 k + b_1 \notin S(a_2 k + b_2) = \{a_2 k + b_2, b_2 k - a_2, n - a_2 k - b_2, n - b_2 k + a_2\}$. As $(a_2 - a_1)k > k$ and $(b_1 - b_2) < k$, we have $a_1 k + b_1 \ne a_2 k + b_2$. Also $(b_2 - a_1)k \ge 2k$ and $a_2 + b_1 \le \lfloor \frac{k-2}{2} \rfloor + k - (a_1 + 1) \le \frac{3k}{2}$; therefore $a_1 k + b_1 \ne b_2 k - a_2$. Note that

$$\begin{aligned} a_1 k + b_1 + a_2 k + b_2 &\le (a_1 + a_2)k + 2k - 2(a_1 + 1) \\ &\le 2\lfloor \frac{k-2}{2} \rfloor k + 2k - 2(a_1 + 1) \\ &\le k^2 - 2k + 2k - 2(a_1 + 1) < k^2 + 1 = n. \end{aligned}$$

Therefore $a_1 k + b_1 \ne n - (a_2 k + b_2)$. Similarly,

$$a_1 k + b_1 + b_2 k - a_2 \le a_1 k + k - (a_1 + 1) + (k - (a_2 + 1))k - a_2 < k^2 + 1 = n,$$

and so $a_1 k + b_1 \ne n - (b_2 k - a_2)$. Hence $S(a_1 k + b_1) \cap S(a_2 k + b_2) = \emptyset$.

Case (ii) $(a_1 < a_2, b_1 < b_2)$ In this case it is very easy to see that $a_1 k + b_1 \notin S(a_2 k + b_2)$, and hence $S(a_1 k + b_1) \cap S(a_2 k + b_2) = \emptyset$.

Case (iii) $(a_1 = a_2, b_1 < b_2)$ Let $a_1 = a_2 = a$. Obviously $ak + b_1 \ne ak + b_2$. Since $0 \le a \le \lfloor \frac{k-2}{2} \rfloor$ and $a + 1 \le b_1 < b_2 \le k - (a+1)$, we have $(b_2 - a)k \ge 2k > (a + b_1)$. Hence $ak + b_1 \ne b_2 k - a$. Also $2ak + b_1 + b_2 \le k(k - 2) + 2k = k^2 < n$, so $ak + b_1 \ne n - (ak + b_2)$. Finally,

$$b_1 + b_2 k + ak - a \le [k - (a + 1)](k + 1) + ak - a = k^2 - 2a - 1 < k^2 + 1 = n,$$

implies $ak + b_1 \ne n - (b_2 k - a)$. Hence $ak + b_1 \notin S(ak + b_2)$ and $S(a_1 k + b_1) \cap S(a_2 k + b_2) = \emptyset$.

Case (iv) $(a_1 < a_2, b_1 = b_2)$ In this case also $S(a_1 k + b_1) \cap S(a_2 k + b_2) = \emptyset$. This completes the proof. $\square$

Now we are ready to define our point process. We neglect $\{0, n/2\}$ if n is even, and $\{0\}$ if n is odd. Let

$$S = \mathbb{Z}_n - \{0, n/2\},$$

$$T_n = \{(a, b) : 0 \leq a \leq \lfloor \frac{k-2}{2} \rfloor, a + 1 \leq b \leq k - (a+1)\}, \qquad (8.18)$$

$$\lambda_t(x) = \frac{1}{\sqrt{n}} \sum_{j=0}^{n-1} x_j \exp(\frac{2\pi j t}{n}),$$

$$\beta_{x,n}(a, b) = \prod_{t \in S(ak+b)} \lambda_t(x), \text{ and}$$

$$\lambda_x(a, b) = (\beta_{x,n}(a, b))^{1/4}.$$

Now define

$$\eta_n(\cdot) = \sum_{(a,b) \in T_n} \epsilon_{\left(\frac{a}{\sqrt{n}}, \frac{b}{\sqrt{n}}, \frac{\lambda_x(a,b) - d_q}{c_q} \right)}(\cdot), \qquad (8.19)$$

where $q = q(n) = \lfloor \frac{n}{4} \rfloor$,

$$c_n = (8 \ln n)^{-1/2}, \text{ and}$$

$$d_n = \frac{(\ln n)^{1/2}}{\sqrt{2}} \left(1 + \frac{1}{4} \frac{\ln \ln n}{\ln n} \right) + \frac{1}{2(8 \ln n)^{1/2}} \ln \frac{\pi}{2}.$$

Theorem 8.4.2 (Bose et al. (2011b)). Let $\{x_t\}$ be i.i.d. random variables which satisfy Assumption 8.2.1. Then $\eta_n \overset{\mathcal{V}}{\to} \eta$, where η is a Poisson point process on $[0, 1/2] \times [0, 1] \times [0, \infty]$ with intensity function $\lambda(s, t, x) = 4\mathbb{I}_{\{s \leq t \leq 1-s\}} e^{-x}$.

Proof. Though the main idea of the proof is similar to the proof of Theorem 8.2.1, the details are more complicated. We do it in two steps.

Step 1: Define

$$\eta_n^*(\cdot) = \sum_{(a,b) \in T_n} \epsilon_{\left(\frac{a}{\sqrt{n}}, \frac{b}{\sqrt{n}}, \frac{\lambda_{\bar{x}} + \sigma_n N(a,b) - d_q}{c_q} \right)}(\cdot).$$

We show $\eta_n^* \overset{\mathcal{D}}{\to} \eta$. Observe that the first two components of the limit are uniformly distributed over a triangle whose vertices are $(0, 0), (1/2, 1/2), (0, 1)$. Denote this triangle by $\triangle$. Since the limit process is simple, it suffices to show that

$$E\eta_n^*((a_1, b_1] \times (a_2, b_2] \times (x, y]) \to E\eta((a_1, b_1] \times (a_2, b_2] \times (x, y]), \qquad (8.20)$$

for all $0 \leq a_1 < b_1 \leq 1/2$, $0 \leq a_2 < b_2 \leq 1$ and $x < y$, and for all $l \geq 1$,

$$P(\eta_n^*((a_1, b_1] \times (c_1, d_1] \times R_1) = 0, \ldots, \eta_n^*((a_l, b_l] \times (c_l, d_l] \times R_l) = 0) \qquad (8.21)$$

$$\longrightarrow P(\eta((a_1, b_1] \times (c_1, d_1] \times R_1) = 0, \dots, \eta((a_l, b_l] \times (c_l, d_l] \times R_l) = 0),$$

where $\cap_{i=1}^{l}(a_i, b_i] \times (c_i, d_i] = \emptyset$ and $R_1, \dots, R_l$ are bounded Borel sets, each consisting of a finite union of intervals on $[0, \infty]$.

Proof of (8.20): Consider the following box sets. See also Figure 8.1.

(i) $(a_1, b_1] \times (a_2, b_2]$ lies entirely inside the triangle $\triangle$.

(ii) $(a_1, b_1] \times (a_1, b_1]$ where $0 \le a_1 < b_1 \le 1/2$.

(iii) $(a_1, b_1] \times (1 - b_1, 1 - a_1]$ where $0 \le a_1 < b_1 \le 1/2$.

(iv) $(a_1, b_1] \times (a_2, b_2]$ lies entirely outside the triangle $\triangle$.

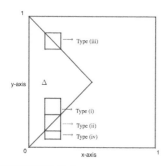 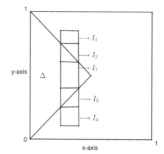

FIGURE 8.1
The left figure shows four types of basic sets and the right figure shows the decomposition of a rectangle into these four types of sets.

Since any rectangle in $[0, 1/2] \times [0, 1]$ can be expressed as the disjoint union of these four kinds of sets (see Figure 8.1), it is sufficient to prove (8.20) and (8.21) for the above four kinds of boxes only. Let I_i denote the i-th type of set. It is enough to prove that for each i, as $n \to \infty$, $E\eta_n^*(I_i \times (x, y]) \to E\eta(I_i \times (x, y])$.

(a) Proof of (8.20) for Type (i) sets:

$$E\eta_n^*((a_1, b_1] \times (a_2, b_2] \times (x, y])$$

$$= E\left(\sum_{(a,b)\in T_n} \epsilon_{\left(\frac{a}{\sqrt{n}}, \frac{b}{\sqrt{n}}, \frac{\lambda_{\bar{x}+\sigma_n N(a,b)} - d_q}{c_q}\right)}((a_1, b_1] \times (a_2, b_2] \times (x, y])\right)$$

$$= \sum_{\left(\frac{a}{\sqrt{n}}, \frac{b}{\sqrt{n}}\right)\in(a_1,b_1]\times(a_2,b_2]} P\left(\frac{\lambda_{\bar{x}+\sigma_n N}(a, b) - d_q}{c_q}\right) \in (x, y])$$

$$\sim (b_1 - a_1)(b_2 - a_2)n\frac{1}{q}(e^{-x} - e^{-y})(1 + o(1))$$

$$\to 4(b_1 - a_1)(b_2 - a_2)(e^{-x} - e^{-y})$$

$$= E\eta((a_1, b_1] \times (a_2, b_2] \times (x, y]).$$

(b) Proof of (8.20) for Type (ii) sets: Similarly,

$$\mathrm{E}\,\eta_n^*((a_1, b_1] \times (a_1, b_1] \times (x, y])$$

$$\sim \frac{1}{2}(b_1 - a_1)(b_1 - a_1)n\frac{1}{q}(e^{-x} - e^{-y})(1 + o(1))$$

$$\to \frac{1}{2}(b_1 - a_1)^2 4(e^{-x} - e^{-y})$$

$$= \mathrm{E}\,\eta((a_1, b_1] \times (a_1, b_1] \times (x, y]).$$

(c) Proof of (8.20) for Type (iii) sets is same as for Type (ii) sets.

(d) Proof of (8.20) for Type (iv) sets:

$$\mathrm{E}\,\eta_n^*((a_1, b_1] \times (a_2, b_2] \times (x, y])$$

$$= \sum_{(\frac{a}{\sqrt{n}}, \frac{b}{\sqrt{n}}) \in (a_1, b_1] \times (a_2, b_2]} \mathrm{P}\,\Big(\frac{\lambda_{\bar{x} + \sigma_n N}(a, b) - d_q}{c_q}\Big) \in (x, y])$$

$$= 0$$

$$= \mathrm{E}\,\eta((a_1, b_1] \times (a_2, b_2] \times (x, y]),$$

since $\{(a, b) \in T_n : (\frac{a}{\sqrt{n}}, \frac{b}{\sqrt{n}}) \in (a_1, b_1] \times (a_2, b_2]\} = \emptyset$. This completes the proof of (8.20).

Proof of (8.21): We prove (8.21) for the four types of sets separately.

(a) Type (i) sets: $(a_i, b_i] \times (c_i, d_i]$ lies completely inside the triangle $\triangle$ for all $i = 1, 2, \ldots, l$. Let

$$n_j = \#\{(a, b) : (\frac{a}{\sqrt{n}}, \frac{b}{\sqrt{n}}) \in (a_j, b_j] \times (c_j, d_j]\}$$

$$\sim \sqrt{n}(b_j - a_j)\sqrt{n}(d_j - c_j)$$

$$= n(b_j - a_j)(d_j - c_j).$$

Then the complement of the event in (8.21) is the union of $m = n_1 + \cdots + n_l$ events, that is,

$$1 - \mathrm{P}\,\big(\eta_n^*((a_1, b_1] \times (c_1, d_1] \times R_1) = 0, \ldots, \eta_n^*((a_l, b_l] \times (c_l, d_l] \times R_l) = 0\big)$$

$$= \mathrm{P}\,\Big(\cup_{j=1}^l \cup_{(\frac{a}{\sqrt{n}}, \frac{b}{\sqrt{n}}) \in (a_j, b_j] \times (c_j, d_j]}\{\frac{\lambda_{\bar{x} + \sigma_n N} - d_q}{c_q} \in R_j\}\Big).$$

Now following the argument to prove (8.6) of Theorem 8.2.1, we get

$$\mathrm{P}\,\big(\eta_n^*((a_1, b_1] \times (c_1, d_1] \times R_1) = 0, \ldots, \eta_n^*((a_l, b_l] \times (c_l, d_l] \times R_l) = 0\big)$$

$$\overset{n \to \infty}{\longrightarrow} \exp\{-\sum_{j=1}^l (b_j - a_j)(d_j - c_j)4\lambda(R_j)\}$$

$$= \mathrm{P}(\eta((a_1, b_1] \times (c_1, d_1] \times R_1) = 0, \ldots, \eta((a_l, b_l] \times (c_l, d_l] \times R_l) = 0).$$

This proves (8.21) for Type (i) sets.

(b) Type (ii) sets: Here $c_i = a_i$, $d_i = b_i$, and

$$n_j \;\; = \;\; \#\{(a,b) : (\frac{a}{\sqrt{n}}, \frac{b}{\sqrt{n}}) \in (a_j, b_j] \times (a_j, b_j]\}$$

$$\sim \frac{1}{2}\sqrt{n}(b_j - a_j)\sqrt{n}(b_j - a_j) = \frac{n}{2}(b_j - a_j)^2.$$

Remaining part of the proof is as in the previous case. Finally we get

$$P\left(\eta_n^*((a_1, b_1] \times (a_1, b_1] \times R_1) = 0, \dots, \eta_n^*((a_l, b_l] \times (a_l, b_l] \times R_l) = 0\right)$$

$$\overset{n \to \infty}{\longrightarrow} \exp\{-\sum_{j=1}^{l} \frac{1}{2}(b_j - a_j)^2 4\lambda(R_j)\}$$

$$= P(\eta((a_1, b_1] \times (a_1, b_1] \times R_1) = 0, \dots, \eta((a_l, b_l] \times (a_l, b_l] \times R_l) = 0).$$

(c) Proof of (8.21) for Type (iii) sets is the same as that given for Type (ii) sets.

(d) For Type (iv) sets, $(a_i, b_i] \times (c_i, d_i] \bigcap \triangle = \emptyset$ for all $i = 1, \dots, l$. Note that for all i,

$$\#\{(a,b) \in T_n : (\frac{a}{\sqrt{n}}, \frac{b}{\sqrt{n}}) \in (a_i, b_i] \times (c_i, d_i]\} = 0,$$

and therefore

$$P\left(\eta_n^*((a_1, b_1] \times (c_1, d_1] \times R_1) = 0, \dots, \eta_n^*((a_l, b_l] \times (c_l, d_l] \times R_l) = 0\right) = 1.$$

Also for the Poisson point process η,

$$P(\eta((a_1, b_1] \times (c_1, d_1] \times R_1) = 0, \dots, \eta((a_l, b_l] \times (c_l, d_l] \times R_l) = 0) = 1.$$

Consequently, the proof of Step 1 is complete.

Step 2: Now define the process,

$$\bar{\eta}_n(\cdot) = \sum_{(a,b) \in T_n} \epsilon_{\left(\frac{a}{\sqrt{n}}, \frac{b}{\sqrt{n}}, \frac{\lambda_{\bar{x}}(a,b) - d_q}{c_q}\right)}(\cdot).$$

Then it suffices to show that for any continuous function f on $[0, 1/2] \times [0, 1] \times [0, \infty)$ with compact support,

$$\bar{\eta}_n(f) - \eta_n^*(f) \overset{P}{\longrightarrow} 0 \text{ and } \bar{\eta}_n(f) - \eta_n(f) \overset{P}{\longrightarrow} 0. \qquad (8.22)$$

Suppose the compact support of f is contained in the set $[0, 1/2] \times [0, 1] \times [K + \gamma_0, \infty)$ for some $\gamma_0 > 0$ and $K \in \mathbb{R}$. Since f is uniformly continuous, as $\gamma \to 0$,

$$\omega(\gamma) := \sup\{|f(s, t, x) - f(s, t, y)|; \; s \in [0, 1/2], \; t \in [0, 1], \; |x - y| \le \gamma\} \to 0.$$

Proof of $\bar{\eta}_n(f) - \eta_n^*(f) \xrightarrow{P} 0$: On the set

$$A_n = \{ \max_{(a,b)\in T_n} |\frac{\lambda_{\bar{x}+\sigma_n N}(a,b)}{c_q} - \frac{\lambda_{\bar{x}}(a,b)}{c_q}| \le \gamma\},$$

we have for $\gamma < \gamma_0$,

$$|f(\frac{a}{\sqrt{n}},\frac{b}{\sqrt{n}},\frac{\lambda_{\bar{x}+\sigma_n N}(a,b)-d_q}{c_q}) - f(\frac{a}{\sqrt{n}},\frac{b}{\sqrt{n}},\frac{\lambda_{\bar{x}}(a,b)-d_q}{c_q})|$$

$$\le \begin{cases} \omega(\gamma) & \text{if } \frac{\lambda_{n,\bar{x}+\sigma_n N}(\omega_j)-b_q}{a_q} > K \\ 0 & \text{if } \frac{\lambda_{n,\bar{x}+\sigma_n N}(\omega_j)-b_q}{a_q} \le K. \end{cases} \tag{8.23}$$

Now if $P(A_n^c) \to 0$, then using (8.23),

$$\limsup_{n\to\infty} P\left(|\eta_n^*(f) - \bar{\eta}_n(f)| > \epsilon\right) \le \frac{\omega(\gamma)}{\epsilon} 4e^{-K} \to 0, \quad \text{as } \gamma \to 0.$$

Now we show that $P(A_n^c) \to 0$. For any random variables $(X_i)_{0\le i < n}$, define

$$M_n(X) = \max_{1\le t\le n} |\frac{1}{\sqrt{n}}\sum_{l=0}^{n-1} X_l \exp(i2\pi tl/n)|.$$

We can use the basic inequalities,

$$\left||z_1 z_2| - |w_1 w_2|\right| \le (|z_1| + |w_2|)\max\{|z_1 - w_1|, |z_2 - w_2|\}, \tag{8.24}$$

and

$$\left||w_1|^{1/2} - |w_2|^{1/2}\right| \le |w_1 - w_2|^{1/2}, \quad z_i, w_i \in \mathbb{C},\ 1 \le i \le 2, \tag{8.25}$$

to obtain,

$$\max_{a,b} \left|\lambda_{\bar{x}+\sigma_n N}(a,b) - \lambda_{\bar{x}}(a,b)\right|$$

$$\le \left[(M_n(\bar{x} + \sigma_n N))^{1/2} + (M_n(\bar{x}))^{1/2}\right](M_n(\sigma_n N))^{1/2} \quad \text{(by (8.24))}$$

$$\le \left[2(M_n(\bar{x} + \sigma_n N))^{1/2} + (M_n(\sigma_n N))^{1/2}\right](M_n(\sigma_n N))^{1/2} \quad \text{(by (8.25))}.$$

By Theorem 2.1 of Davis and Mikosch (1999) (see Theorem 11.3.2 in Appendix), we have

$$M_n^2(\sigma_n N) = O_p(\sigma_n^2 \ln n) \quad \text{and} \quad M_n^2(\bar{x} + \sigma_n N) = O_p(\ln n),$$

with $\sigma_n^2 = n^{-c}$. Therefore

$$\max_{a,b} \frac{1}{c_q}\left|\lambda_{\bar{x}+\sigma_n N}(a,b) - \lambda_{\bar{x}}(a,b)\right| = O_p((\ln n)n^{-c/2}).$$

Hence

$$P(A_n^c) = P\left(\max_{a,b}\left|\frac{\lambda_{\bar{x}+\sigma_n N}(a,b)}{c_q} - \frac{\lambda_{\bar{x}}(a,b)}{c_q}\right| > \epsilon\right)$$

$$= P\left(\frac{n^{c/4}}{\ln n}\max_{a,b}\frac{1}{c_q}\left|\lambda_{\bar{x}+\sigma_n N}(a,b) - \lambda_{\bar{x}}(a,b)\right| > \frac{\epsilon n^{c/4}}{\ln n}\right) \to 0 \text{ as } n \to \infty.$$

Hence $\bar{\eta}_n(f) - \eta_n^*(f) \xrightarrow{P} 0$. The other part of (8.22) follows from (6.14) of Chapter 6. Proof of Step 2 and hence of the theorem is now complete. $\square$

We invite the reader to formulate and prove a result similar to Corollary 8.2.2 for the ordered eigenvalues of the k-circulant matrix.

8.5   Reverse circulant: dependent input

Let $\{x_n; n \geq 0\}$ be a two sided moving average process,

$$x_n = \sum_{i=-\infty}^{\infty} a_i \varepsilon_{n-i}, \tag{8.26}$$

where $\{a_n\}$ are non-random with $\sum_n |a_n| < \infty$, and $\{\varepsilon_i; i \in \mathbb{Z}\}$ are i.i.d. random variables. Let $f(\omega)$, $\omega \in [0, 2\pi]$ be the spectral density of $\{x_n\}$.

Assumption 8.5.1. $\{\varepsilon_i; i \in \mathbb{Z}\}$ are i.i.d. random variables with $\mathrm{E}(\varepsilon_0) = 0$, $\mathrm{E}(\varepsilon_0^2) = 1$, and $\mathrm{E}|\varepsilon_0|^s < \infty$ for some $s > 2$,

$$\sum_{j=-\infty}^{\infty} |a_j||j|^{1/2} < \infty, \text{ and } f(\omega) > 0 \text{ for all } \omega \in [0, 2\pi].$$

Let $\lambda_{n,x}(\omega_k)$ be the eigenvalues of $\frac{1}{\sqrt{n}}RC_n$ as given in (8.1). As in Chapter 7, we scale each eigenvalue and consider the point process,

$$\tilde{\eta}_n(\cdot) = \sum_{j=1}^{q} \epsilon_{\left(\omega_j, \frac{\tilde{\lambda}_{n,x}(\omega_k) - b_q}{a_q}\right)}(\cdot), \tag{8.27}$$

where $\tilde{\lambda}_{n,x}(\omega_j) = \frac{\lambda_{n,x}(\omega_j)}{\sqrt{2\pi f(\omega_j)}}$, $a_q = \frac{1}{2\sqrt{\ln q}}$, $b_q = \sqrt{\ln q}$ and $q = q(n) = \lfloor \frac{n}{2} \rfloor$.

Theorem 8.5.1 (Bose et al. (2011b)). Let $\{x_n\}$ be as in (8.26) and which satisfies Assumption 8.5.1. Then $\tilde{\eta}_n \xrightarrow{\mathcal{V}} \eta$, where η is a Poisson point process on $[0, \pi] \times (-\infty, \infty]$ with intensity function $\lambda(t, x) = \pi^{-1}e^{-x}$.

Proof. Define

$$\eta_n(\cdot) = \sum_{j=1}^{q} \epsilon_{\left(\omega_j, \frac{\lambda_{n,\varepsilon}(\omega_j) - b_q}{a_q}\right)}(\cdot). \tag{8.28}$$

In Theorem 8.2.1, we have shown that $\eta_n \overset{\mathcal{V}}{\to} \eta$. Now it is enough to show that for any continuous function g on E with compact support,

$$\tilde{\eta}_n(g) - \eta_n(g) \overset{P}{\to} 0 \quad \text{as} \quad n \to \infty.$$

Suppose the compact support of g is contained in the set $[0, \pi] \times [K + \gamma_0, \infty)$ for some $\gamma_0 > 0$ and $K \in \mathbb{R}$. Since g is uniformly continuous,

$$\omega(\gamma) := \sup\{|g(t, x) - g(t, y)|; \ t \in [0, 1], \ |x - y| \le \gamma\} \to 0 \text{ as } \gamma \to 0.$$

On the set

$$A_n = \{ \max_{j=1,\dots,q} |\frac{\lambda_{n,x}(\omega_j)}{a_q \sqrt{2\pi f(\omega_j)}} - \frac{\lambda_{n,\varepsilon}(\omega_j)}{a_q}| \le \gamma\},$$

we have for $\gamma < \gamma_0$,

$$|g(\omega_j, \frac{\tilde{\lambda}_{n,x}(\omega_j) - b_q}{a_q}) - g(\omega_j, \frac{\lambda_{n,\varepsilon}(\omega_j) - b_q}{a_q})| \le \begin{cases} \omega(\gamma) & \text{if } \frac{\lambda_{n,\varepsilon}(\omega_j) - b_q}{a_q} > K \\ 0 & \text{if } \frac{\lambda_{n,\varepsilon}(\omega_j) - b_q}{a_q} \le K. \end{cases} \tag{8.29}$$

Observe that

$$\frac{1}{a_q} \max_{1 \le j \le q} |\frac{\lambda_{n,x}(\omega_j)}{\sqrt{2\pi f(\omega_j)}} - \lambda_{n,\varepsilon}(\omega_j)|$$

$$\le \frac{1}{\alpha a_q} \max_{1 \le j \le q} |\lambda_{n,x}(\omega_j) - \sqrt{2\pi f(\omega_j)}\lambda_{n,\varepsilon}(\omega_j)|$$

$$\le \frac{1}{\alpha a_q} \max_{1 \le j \le q} |\frac{1}{\sqrt{n}} \sum_{l=0}^{n-1} x_l e^{i\omega_j l} - (\sum_{t=-\infty}^{\infty} a_t e^{i\omega_j t}) \frac{1}{\sqrt{n}} \sum_{l=0}^{n-1} \varepsilon_l e^{i\omega_j l}|$$

$$= o_p(n^{-1/4}), \text{ by (7.5) of Chapter 7.}$$

Therefore $\lim_{n \to \infty} P(A_n^c) = 0$. Now, for any $\delta > 0$, choose $0 < \gamma < \gamma_0$. Then, by intersecting the event $\{|\tilde{\eta}_n(g) - \eta_n(g)| > \delta\}$ with A_n and A_n^c and using (8.29), we obtain

$$\limsup_{n \to \infty} P(|\tilde{\eta}_n(g) - \eta_n(g)| > \delta)$$

$$\le \limsup_{n \to \infty} (P(\{|\tilde{\eta}_n(g) - \eta_n(g)| > \delta\} \cap A_n) + P(A_n^c))$$

$$\le \limsup_{n \to \infty} P(\omega(\gamma)\eta_n([0, \pi] \times [K, \infty)) > \epsilon) + \limsup_{n \to \infty} P(A_n^c)$$

$$\le \limsup_{n \to \infty} E \eta_n([0, \pi] \times [K, \infty))\omega(\gamma)/\epsilon \le e^{-K}\omega(\gamma)/\epsilon.$$

Since $\omega(\gamma) \to 0$ as $\gamma \to 0$, $\tilde{\eta}_n - \eta_n \overset{P}{\to} 0$. $\qquad\square$

8.6 Symmetric circulant: dependent input

Now we consider the two sided moving average process defined in (8.26) with an extra assumption that $a_j = a_{-j}$ for all $j \in \mathbb{N}$. Define

$$\tilde{\eta}_n(\cdot) = \sum_{j=0}^{q} \epsilon_{\left(\omega_j, \frac{\tilde{\lambda}_{n,x}(\omega_j) - b_q}{a_q}\right)}(\cdot), \tag{8.30}$$

where $q = q(n) \sim \frac{n}{2}$, $\tilde{\lambda}_{n,x}(\omega_j) = \frac{\lambda_{n,x}(\omega_j)}{\sqrt{2\pi f(\omega_j)}}$, and $\lambda_{n,x}(\omega_j)$ are the eigenvalues of the symmetric circulant matrix given in (8.14), and a_q, b_q are as in (8.16).

Theorem 8.6.1 (Bose et al. (2011b)). Let $\{x_n\}$ be as in (8.26) with $a_j = a_{-j}$ and which satisfies Assumption 8.5.1. Then $\tilde{\eta}_n \xrightarrow{\mathcal{V}} \eta$, where η is a Poisson point process on $[0, \pi] \times (-\infty, \infty]$ with intensity function $\lambda(t, x) = \pi^{-1} e^{-x}$.

Proof. The proof is very similar to the proof of Theorem 8.5.1. We only point out that to show $\lim_{n \to \infty} \mathrm{P}(A_n^c) = 0$, we use the following fact from (7.12) of Chapter 7:

$$\max_{1 \le k \le \lfloor \frac{n}{2} \rfloor} \left| \frac{\lambda_{n,x}(\omega_k)}{\sqrt{2\pi f(\omega_k)}} - \lambda_{n,\varepsilon}(\omega_k) \right| = o_p(n^{-1/4}).$$

$\square$

8.7 k-circulant, $n = k^2 + 1$: dependent input

First recall the eigenvalues of the k-circulant matrix for $n = k^2 + 1$ given in Section (8.4) and define the following notation based on that:

$$\beta_{\varepsilon,n}(a, b) = \prod_{t \in S(ak+b)} \lambda_t(\varepsilon), \quad \lambda_\varepsilon(a, b) = (\beta_{\varepsilon,n}(a, b))^{1/4},$$

$$\tilde{\beta}_{x,n}(a, b) = \frac{\prod_{t \in S(ak+b)} \lambda_t(x)}{4\pi^2 f(\omega_{ak+b}) f(\omega_{bk-a})}, \quad \text{and} \quad \tilde{\lambda}_x(a, b) = (\tilde{\beta}_{x,n}(a, b))^{1/4}.$$

Now with $q = q(n) = \lfloor \frac{n}{4} \rfloor$, and d_q, c_q as in (8.20), define our point process based on points $\{(\frac{a}{\sqrt{n}}, \frac{b}{\sqrt{n}}, \frac{\tilde{\lambda}_x(a,b) - d_q}{c_q}) : (a, b) \in T_n\}$ as:

$$\tilde{\eta}_n(\cdot) = \sum_{(a,b) \in T_n} \epsilon_{\left(\frac{a}{\sqrt{n}}, \frac{b}{\sqrt{n}}, \frac{\tilde{\lambda}_x(a,b) - d_q}{c_q}\right)}(\cdot), \tag{8.31}$$

where T_n is as defined in (8.18).

Theorem 8.7.1 (Bose et al. (2011b)). Let $\{x_n\}$ be as in (8.26) and which satisfies Assumption 8.5.1. Then $\tilde{\eta}_n \overset{\mathcal{V}}{\to} \eta$, where η is a Poisson point process on $[0, 1/2] \times [0, 1] \times [0, \infty]$ with intensity function $\lambda(s, t, x) = 4\mathbb{I}_{\{s \leq t \leq 1-s\}}e^{-x}$.

Proof. First define a point process based on $\{(\frac{a}{\sqrt{n}}, \frac{b}{\sqrt{n}}, \frac{\lambda_\varepsilon(a,b)-d_q}{c_q}) : (a, b) \in T_n\}$. In Theorem 8.4.2, we have shown that $\eta_n \overset{\mathcal{D}}{\to} \eta$, where η is a Poisson point process on $[0, 1/2] \times [0, 1] \times (-\infty, \infty]$ with intensity function $\lambda(s, t, x) = 4\mathbb{I}_{\{s \leq t \leq 1-s\}}e^{-x}$. The rest of the argument is similar to that in the proof of Theorem 8.5.1. The additional fact that is needed is:

$$P(\max_{(a,b) \in T_n} \left| \tilde{\lambda}_x(a, b) - \lambda_\varepsilon(a, b) \right| > \gamma) \to 0,$$

which follows from the proof of Theorem 7.4.1 of Chapter 7.   □

8.8   Exercises

1. Complete the proof of (8.10).

2. Complete the proof of Corollary 8.3.2.

3. Complete the proof of Theorem 8.6.1.

4. Complete the proof of Theorem 8.7.1.

9

Heavy-tailed input: LSD

In this chapter we consider the circulant-type matrices when the input sequence belongs to the domain of attraction of an α-stable law with $\alpha \in (0, 2)$. We show that the LSDs are random distributions in these cases and determine explicit representations of the limits. The method of proof relies heavily on the ideas borrowed from the results available outside the random matrix context.

9.1   Stable distribution and input sequence

A random variable Y_α is said to have a *stable distribution* $S_\alpha(\sigma, \beta, \mu)$ if there are parameters $0 < \alpha \leq 2, \sigma \geq 0, -1 \leq \beta \leq 1$ and μ real such that its characteristic function has the following form:

$$
E[\exp(itY_\alpha)] = \begin{cases} \exp\{i\mu t - \sigma^\alpha |t|^\alpha (1 - i\beta \operatorname{sgn}(t) \tan(\pi\alpha/2))\}, & \text{if } \alpha \neq 1, \\[2mm] \exp\{i\mu t - \sigma |t|(1 + (2i\beta/\pi) \operatorname{sgn}(t) \ln |t|)\}, & \text{if } \alpha = 1. \end{cases}
$$

An excellent reference for stable distributions and stable processes is Samorodnitsky and Taqqu (1994).

A sequence of i.i.d. random variables $\{X_i\}$ is said to belong to the domain of attraction of a stable law with index α if there exists $a_n \to \infty$ such that

$$
\frac{1}{a_n} \sum_{k=1}^{n} (X_k - c_n) \xrightarrow{\mathcal{D}} S_\alpha,
$$

where S_α is a stable random variable and $c_n = E[X_1 \mathbb{I}(|X_1| \leq a_n)]$.

It is well-known that a random variable X is in the domain of attraction of a (non-normal) stable law with index $\alpha \in (0, 2)$ if and only if

$$
P[|X| > t] = t^{-\alpha} \ell(t),
$$

for some *slowly varying function* $\ell(\cdot)$ and

$$
\lim_{t \to \infty} \frac{P[X > t]}{P[|X| > t]} = p \in [0, 1]. \tag{9.1}
$$

In that case, the normalizing constants a_n can be chosen so that

$$n\,\mathrm{P}[|X| > a_n x] \to x^{-\alpha}. \tag{9.2}$$

In this chapter we deal with the situation where the elements of the input sequence are in the domain of attraction of a stable distribution. Thus, we make the following assumption on the input sequence $\{X_i\}$.

Assumption 9.1.1. The elements of the input sequence $\{X_i\}$ are i.i.d. random variables in the domain of attraction of a stable law with index $\alpha \in (0, 2)$.

9.2   Background

So far, the LSDs, either in probability or almost surely, have all been non-random. But with heavy-tailed entries this is not the case anymore. To understand the mode of convergence of ESDs of circulant-type matrices with heavy-tailed entries, we need to first discuss the convergence of a sequence of random probability measures to a random probability measure.

Let $(\Omega, \mathcal{A}, \mathrm{P})$ be a probability space and μ be a measurable map from $(\Omega, \mathcal{A}, \mathrm{P})$ to the set of probability measures on $\mathbb{R}$. Hence $\mu(\cdot, \omega)$ is a probability measure on $\mathbb{R}$. Let $\{\mu_n\}$ be a sequence of measurable maps from $(\Omega, \mathcal{A}, \mathrm{P})$ to the set of probability measures on $\mathbb{R}$. We say μ_n converges to the random measure μ weakly if for any bounded continuous function $f : \mathbb{R} \to \mathbb{R}$,

$$\int f(x)\mu_n(dx, \omega) \to \int f(x)\mu(dx, \omega), \quad \text{for almost every } \omega \in \Omega.$$

Note that $Y_n(\omega) := \int f(x)\mu_n(dx, \omega)$ and $Y(\omega) := \int f(x)\mu_n(dx, \omega)$ are real-valued random variables defined on $(\Omega, \mathcal{A}, \mathrm{P})$. Hence the above weak convergence is the almost sure convergence of the random variables Y_n to Y.

Also recall Skorohod's representation theorem which lifts weak convergence to almost sure convergence. This result will be a key ingredient in the proofs of the results of this chapter.

Skorohod's representation theorem: Let $\{\mu_n\}$ be a probability measure on a completely separable metric space S such that μ_n converges weakly to some probability measure μ on S as $n \to \infty$. Then there exists a sequence of S-valued random variables $\{X_n\}$ and another random variable X, all defined on a common probability space $(\Omega, \mathcal{A}, \mathrm{P})$ such that the law of X_n is μ_n for all n, the law of X is μ, and X_n converges to X almost surely.

A few results from Knight (1991) will be of direct use in our situation. We describe these now. Suppose $\{X_i\}$ are i.i.d. random variables in the domain of attraction of a stable law with index $\alpha \in (0, 2)$. Then, following Knight (1991), order the X_k's, $1 \le k \le n$ by their magnitudes $|X_{n1}| \ge |X_{n2}| \ge \cdots \ge |X_{nn}|$.

Let $Y_{nk} = |X_{nk}|$ and B_{nk} be the random signs so that $X_{nk} = B_{nk}Y_{nk}$. Let $\pi_{n1}, \pi_{n2}, \ldots, \pi_{nn}$ be the anti-ranks of $Y_{n1}, Y_{n2}, \ldots, Y_{nn}$, that is, for $1 \le k \le n$,

$$\pi_{nk}(\omega) = l \quad \text{if} \quad Y_{nk}(\omega) = |X_l(\omega)| \quad \text{for} \quad \omega \in \Omega.$$

Define

$$U_{nk} = \frac{\pi_{nk}}{n} \quad \text{and} \quad Z_{nk} = \frac{1}{a_n}Y_{nk}.$$

Suppose $\{U_k\}$ are i.i.d. random variables that are uniformly distributed on $(0,1)$. The variables $\{B_k\}$ are i.i.d. random variables with $\mathrm{P}[B_1 = 1] = p = 1 - \mathrm{P}[B_1 = -1]$ where p is defined by equation (9.1). The sequence $\{Z_k\}$ is defined as

$$Z_k = \Gamma_k^{-1/\alpha} = \Big(\sum_{i=1}^{k} E_i\Big)^{-1/\alpha},$$

where $\{E_j\}$ are i.i.d. exponential random variables with mean 1. Now we have the following lemmata from Knight (1991).

Lemma 9.2.1 (Knight (1991), Lemma 1). In $\mathbb{R}^\infty$, we have

$$\boldsymbol{U}^n = (U_{n1}, \ldots, U_{nn}, 0, 0, \ldots) \xrightarrow{\mathcal{D}} \boldsymbol{U} = (U_1, U_2, \ldots),$$

$$\boldsymbol{B}^n = (B_{n1}, \ldots, B_{nn}, 0, 0, \ldots) \xrightarrow{\mathcal{D}} \boldsymbol{B} = (B_1, B_2, \ldots),$$

$$\boldsymbol{Z}^n = (Z_{n1}, \ldots, Z_{nn}, 0, 0, \ldots) \xrightarrow{\mathcal{D}} \boldsymbol{Z} = (Z_1, Z_2, \ldots).$$

Moreover, in the above, the three sequences converge jointly with the three limiting sequences being mutually independent.

Now due to Skorohod's representation theorem, we can assume that all the above random variables are defined on a suitable probability space $(\Omega, \mathcal{A}, \mathrm{P})$ where the above convergence in distribution holds almost surely. That is, $\{U_{nk}\}$, $\{B_{nk}\}$, $\{Z_{nk}\}$, $\{U_k\}$, $\{B_k\}$ and $\{Z_k\}$ are defined on the probability space $(\Omega, \mathcal{A}, \mathrm{P})$ and $\boldsymbol{U}^n \to \boldsymbol{U}$, $\boldsymbol{B}^n \to \boldsymbol{B}$ and $\boldsymbol{Z}^n \to \boldsymbol{Z}$ almost surely. A typical element of Ω will be denoted by ω.

Lemma 9.2.2 (Knight (1991), Lemma 2). Suppose that the convergence results of Lemma 9.2.1 hold in almost sure sense and let $0 < \varepsilon < \gamma \le \infty$.

(a) Then, as $n \to \infty$,

$$\sum_{k=1}^{\infty} \big|B_{nk}Z_{nk}\mathbb{I}(\varepsilon < Z_{nk} \le \gamma) - B_k Z_k \mathbb{I}(\varepsilon < Z_k \le \gamma)\big| \to 0, \quad \text{almost surely.}$$

(b) If $\gamma < \infty$ or if $\gamma \le \infty$ and $\alpha > 1$ then, as $n \to \infty$,

$$\sum_{k=1}^{\infty} \mathrm{E}\,\big(\big|B_{nk}Z_{nk}\mathbb{I}(\varepsilon < Z_{nk} \le \gamma) - B_k Z_k \mathbb{I}(\varepsilon < Z_k \le \gamma)\big|\big) \to 0.$$

Now we state another important lemma from Knight (1991). We need the following notion of rational independence to state the result.

Definition 9.2.1. A sequence of real numbers $\{b_j\}_{1 \leq j \leq m} \in (0,1)$ is said to be *rationally independent* if for integers $\{a_j\}_{1 \leq j \leq m}$,

$$a_1 b_1 + \cdots + a_m b_m = 0 \;(\mathrm{mod}\; 1) \quad \Leftrightarrow \quad a_1 = \cdots = a_m = 0.$$

Lemma 9.2.3 (Knight (1991), Lemma 3). Let $\{\pi_{nj}\}_{\{j=1,\ldots,m\}}$, $n \geq 1$ be positive real numbers such that

$$\frac{\pi_{nj}}{n} \to u_j \in (0,1), \tag{9.3}$$

where $\{u_j\}$ are rationally independent. Let $\boldsymbol{R}_n$ be a random vector taking values

$$\left(t\frac{\pi_{n1}}{n}(\mathrm{mod}\; 1),\ldots,t\frac{\pi_{nm}}{n}(\mathrm{mod}\; 1)\right), \;\; t = 1,2,\ldots,n,$$

with probability $1/n$ for each t. Then the distribution of $\boldsymbol{R}_n$ converges weakly to the uniform distribution on the unit cube $[0,1]^m$.

Now we state the main result of Knight (1991). The LSD results for the symmetric circulant matrix will follow from it.

Theorem 9.2.4 (Knight (1991), Theorem 5). Let $h(\cdot)$ be a continuous function with period one and $\{X_j\}$ be i.i.d. as in Assumption 9.1.1. Define

$$\hat{X}_j(h) = \frac{1}{a_n}\sum_{k=1}^{n} h(jk/n)X_k \quad \text{for} \quad j = 1,2,\ldots,n;$$

$$\tilde{X}_j(h) = \frac{1}{a_n}\sum_{k=1}^{n} h(jk/n)(X_k - c_n) \quad \text{for} \quad j = 1,2,\ldots,n;$$

$$\hat{\mathrm{P}}_n^*(A) = \frac{1}{n}\sum_{j=1}^{n} \mathbb{I}(\hat{X}_j(h) \in A), \quad \text{empirical measure of } \{\hat{X}_j(h)\};$$

$$\tilde{\mathrm{P}}_n^*(A) = \frac{1}{n}\sum_{j=1}^{n} \mathbb{I}(\tilde{X}_j(h) \in A), \quad \text{empirical measure of } \{\tilde{X}_j(h)\};$$

where a_n is as in (9.2) and $c_n = \mathrm{E}[X_1\mathbb{I}(|X_1| \leq a_n)]$.

(a) If $0 < \alpha < 2$ then

$$\tilde{\mathrm{P}}_n^* \xrightarrow{\mathcal{D}} \tilde{\mathrm{P}}^*,$$

where $\tilde{\mathrm{P}}^*(\cdot,\omega)$ is the probability distribution of

$$\sum_{k=1}^{\infty} h(U_k^*)\big(B_k(\omega)Z_k(\omega) - \mathrm{E}(B_k Z_k \mathbb{I}(Z_k < 1))\big),$$

and $\{U_k^*\}$ is a sequence of i.i.d. uniform random variables on $(0,1)$.

(b) If, either, (i) $\#\{j : \sum_{k=1}^n h(jk/n) \neq 0\} = o(n)$ and $\int_0^1 h(x)dx = 0$, or (ii) $\alpha > 1$ and $E(X_1) = 0$, or (iii) $\alpha < 1$, then

$$\hat{P}_n^* \xrightarrow{\mathcal{D}} \hat{P}^*,$$

where $\hat{P}^*$ is the probability distribution of

$$\sum_{k=1}^\infty h(U_k^*)B_k(\omega)Z_k(\omega).$$

We will not give a detailed proof of Theorem 9.2.4; instead we outline the main idea to prove convergence in distribution of $\hat{P}_n^*$ to $\hat{P}^*$. The idea of the proof of $\tilde{P}_n^* \xrightarrow{\mathcal{D}} \tilde{P}^*$ is similar. We will follow the same idea to prove the LSD results of circulant-type matrices.

First observe that

$$\hat{X}_j(h) = \sum_{k=1}^n h(jU_{nk})B_{nk}Z_{nk} = \sum_{k=1}^\infty h(jU_{nk})B_{nk}Z_{nk},$$

where $\{U_{nk}\}$, $\{B_{nk}\}$ and $\{Z_{nk}\}$ are as defined before. So, sampling from $\hat{P}_n^*$ produces the random variable

$$\sum_{k=1}^\infty h(U_{nk}^*)B_{nk}Z_{nk},$$

where $(U_{n1}^*, U_{n2}^*, \ldots, U_{nn}^*, 0, \ldots)$ is chosen at random from the set of sequences

$$\{(jU_{n1} \pmod 1), \ldots, jU_{nn} \pmod 1), 0, \ldots), \; j = 1, 2, \ldots, n\}.$$

Now from Lemma 9.2.1 and Skorohod's representation theorem, $\{U_{nk}\}$, $\{B_{nk}\}$, $\{Z_{nk}\}$, $\{U_k\}$, $\{B_k\}$ and $\{Z_k\}$ are defined on the same probability space $(\Omega, \mathcal{A}, P)$ and $\boldsymbol{U}^n \to \boldsymbol{U}$, $\boldsymbol{B}^n \to \boldsymbol{B}$ and $\boldsymbol{Z}^n \to \boldsymbol{Z}$ almost surely. Moreover the limiting sequences $\{U_k\}$, $\{B_k\}$ and $\{Z_k\}$ are mutually independent.

Now Lemma 9.2.3 suggests that $(U_{n1}^*, U_{n2}^*, \ldots, U_{nn}^*, 0, \ldots)$ will converge in distribution to $(U_1^*, U_2^*, U_3^*, \ldots)$, and hence $\hat{P}_n^*$ will converge in distribution to the random probability measure of the (random) random variables

$$\sum_{k=1}^\infty h(U_k^*)B_k Z_k,$$

where $\{U_k^*\}$ is a sequence of i.i.d. uniform random variables on $(0,1)$. Note that the randomness of the limiting measure is induced by $\{B_k\}$ and $\{Z_k\}$.

We shall follow a similar idea to prove the LSD results in the following sections.

9.3 Reverse circulant and symmetric circulant

Let $\{\lambda_j\}$ denote the eigenvalues of the random matrix $\frac{1}{a_n}M_n$ with input sequence $\{X_i\}$, where $\{a_n\}$ is a positive sequence of constants as defined in (9.2). Then we denote the ESD of $\frac{1}{a_n}M_n$ by L_{M_n} so that,

$$L_{M_n}(A) = \frac{1}{n}\sum_{j=1}^{n}\mathbb{I}(\lambda_j \in A).$$

If the input sequence for M_n is $\{X_i - c_n\}$ with $c_n = \mathrm{E}[X_1\mathbb{I}(|X_1| \le a_n)]$, then the eigenvalues of $\frac{1}{a_n}M_n$ will be denoted by $\{\tilde{\lambda}_i\}$ and the corresponding ESD will be denoted by $\tilde{L}_{M_n}$. So

$$\tilde{L}_{M_n}(A) = \frac{1}{n}\sum_{j=1}^{n}\mathbb{I}(\tilde{\lambda}_j \in A).$$

Using the methods of Freedman and Lane (1981), Bose et al. (2003) showed that the LSD of the reverse circulant matrix with heavy tailed input exists. However, no closed form representation of the limit was given. On the other hand, Knight (1991) considered the empirical distribution of the periodogram of the $\{X_i\}$ and was able to obtain some very nice representations for it. One can combine these two results to state the following theorem. It also follows as a special case of Theorem 9.4.2 which is stated later.

We recall random variables $\{B_j\}$, $\{\Gamma_j\}$ and $\{Z_j\}$ from Section 9.2 to state the LSD results for the reverse circulant, the symmetric circulant and the k-circulant matrices where the input sequence $\{X_j\}$ satisfies Assumption 9.1.1. $\{B_j\}$ and $\{\Gamma_j\}$ are independent random sequences defined on the same probability space where B_j are i.i.d. with $\mathrm{P}[B_1 = 1] = p = 1 - \mathrm{P}[B_1 = -1]$, p is defined by equation (9.1), $\Gamma_j = \sum_{i=1}^{j}E_i$, $Z_j = \Gamma_j^{-1/\alpha}$, and $\{E_i\}$ are i.i.d. exponential random variables with mean 1. Now define

$$\mu_t = \mathrm{E}[B_t Z_t \mathbb{I}(Z_t \le 1)]. \tag{9.4}$$

Let $\{U_j^*\}$ be a sequence of i.i.d. uniform random variables on $(0,1)$.

Theorem 9.3.1 (Bose et al. (2011a)). (a) $\tilde{L}_{RC_n} \overset{\mathcal{D}}{\to} \tilde{L}_{RC}$, where $\tilde{L}_{RC}(\cdot,\omega)$ is the random distribution of the symmetric square root of

$$\Big(\sum_{t=1}^{\infty}\cos(2\pi U_t^*)(B_t(\omega)Z_t(\omega) - \mu_t)\Big)^2 + \Big(\sum_{t=1}^{\infty}\sin(2\pi U_t^*)(B_t(\omega)Z_t(\omega) - \mu_t)\Big)^2.$$

(b) $L_{RC_n} \overset{\mathcal{D}}{\to} L_{RC}$, where $L_{RC}(\cdot,\omega)$ is the random distribution of the symmetric square root of

$$\Big(\sum_{t=1}^{\infty}\cos(2\pi U_t^*)B_t(\omega)Z_t(\omega)\Big)^2 + \Big(\sum_{t=1}^{\infty}\sin(2\pi U_t^*)B_t(\omega)Z_t(\omega)\Big)^2.$$

We now give the corresponding result for the symmetric circulant matrix.

Theorem 9.3.2 (Bose et al. (2011a)). (a) $\tilde{L}_{SC_n} \overset{D}{\to} \tilde{L}_{SC}$, where $\tilde{L}_{SC}(\cdot, \omega)$ is the distribution of

$$2 \sum_{t=1}^{\infty} \cos(2\pi U_t^*)(B_t(\omega)Z_t(\omega) - \mu_t).$$

(b) $L_{SC_n} \overset{D}{\to} L_{SC}$, where $L_{SC}(\cdot, \omega)$ is the distribution of

$$2 \sum_{t=1}^{\infty} \cos(2\pi U_t^*)B_t(\omega)Z_t(\omega).$$

We briefly sketch a proof of the above result. Let $\lambda_0, \lambda_1, \ldots, \lambda_{n-1}$ be the eigenvalues of $\frac{1}{a_n}SC_n$ given by:

(i) for n odd:

$$\begin{cases} \lambda_0 &= \frac{1}{a_n}\left(X_1 + 2\sum_{j=1}^{[n/2]} X_{j+1}\right) \\ \lambda_k &= \frac{1}{a_n}\left(X_1 + 2\sum_{j=1}^{[n/2]} X_{j+1}\cos(\frac{2\pi jk}{n})\right), \quad 1 \le k \le [n/2]; \end{cases} \quad (9.5)$$

(ii) for n even:

$$\begin{cases} \lambda_0 &= \frac{1}{a_n}\left(X_1 + 2\sum_{j=1}^{\frac{n}{2}-1} X_{j+1} + X_{n/2}\right) \\ \lambda_k &= \frac{1}{a_n}\left(X_1 + 2\sum_{j=1}^{\frac{n}{2}-1} X_{j+1}\cos(\frac{2\pi jk}{n}) + (-1)^k X_{\frac{n}{2}+1}\right), \quad 1 \le k \le \frac{n}{2} \end{cases}$$

with $\lambda_{n-k} = \lambda_k$ in both cases.

Now observe that, as far as determining the LSD of $\frac{1}{a_n}SC_n$ is concerned, we can ignore the quantity $\frac{1}{a_n}X_1$ and also the quantity $\frac{1}{a_n}X_{n/2+1}$, if n is even, from the eigenvalue formula since $\frac{1}{a_n}X_1 \to 0$ in probability. Then the results follow from Theorem 9.2.4.

9.4 *k*-circulant: $n = k^g + 1$

First suppose $n = k^2 + 1$. Then from Lemma 6.2.1, if k is even then there is one singleton partition set $\{0\}$, and if k is odd then there are two singleton partition sets $\{0\}$ and $\{n/2\}$, respectively (which we may ignore); all the remaining partitions have four elements each.

In general, for $n = k^g + 1$, $g \ge 1$, the eigenvalue partition (see Section 6.2 of Chapter 6) of $\{0, 1, 2, \ldots, n-1\}$ contains approximately $q = [\frac{n}{2g}]$ sets each of size $(2g)$ and each set is self-conjugate; in addition, the remaining sets do not contribute to the LSD. We shall call the partition sets of size $(2g)$ the *major partition sets*.

We will benefit from expressing the eigenvalues in a convenient form. This is given in the following lemma for easy reference. To do this, observe that a typical $S(x)$ may be written as

$$S(b_1 k^{g-1} + b_2 k^{g-2} + \cdots + b_g),$$

which in turn is the union of the following two sets:

$$\left\{ b_1 k^{g-1} + b_2 k^{g-2} + \cdots + b_g, b_2 k^{g-1} + b_3 k^{g-2} + \cdots + b_g k - b_1, \ldots, \right.$$
$$\left. b_g k^{g-1} - b_1 k^{g-2} - \cdots - b_{g-1} \right\}$$

and its conjugate, that is,

$$\left\{ n - (b_1 k^{g-1} + b_2 k^{g-2} + \cdots + b_g), \ldots, n - (b_g k^{g-1} - b_1 k^{g-2} - \cdots - b_{g-1}) \right\},$$

where

$$0 \le b_1 \le k - 1, \ \ldots, 0 \le b_{g-1} \le k - 1 \ \text{ and } \ 1 \le b_g \le k.$$

Define

$$T_n = \{ (b_1, b_2, \ldots, b_g) : \ 0 \le b_1 \le k - 1, \ldots, 1 \le b_g \le k \},$$

$$C_t = \sum_{j=1}^{n} X_j \cos\left(\frac{2\pi jt}{n}\right) \quad \text{and} \quad S_t = \sum_{j=1}^{n} X_j \sin\left(\frac{2\pi jt}{n}\right), \quad \text{for } t \in \mathbb{N}.$$

Lemma 9.4.1. The eigenvalues of the k-circulant $a_n^{-1} A_{k,n}$ with $n = k^g + 1$ corresponding to the major partition sets may be written as

$$\left\{ \lambda_{(b_1, b_2, \ldots, b_g)}, \lambda_{(b_1, b_2, \ldots, b_g)} \omega_{2g}, \ldots, \lambda_{(b_1, b_2, \ldots, b_g)} \omega_{2g}^{2g-1} : (b_1, b_2, \ldots, b_g) \in T_n \right\},$$

where ω_{2g} is a primitive $(2g)$-th root of unity and

$$a_n \lambda_{(b_1, b_2, \ldots, b_g)}$$
$$= \left(C_{b_1 k^{g-1} + \cdots + b_g}^2 + S_{b_1 k^{g-1} + \cdots + b_g}^2 \right)^{\frac{1}{2g}} \ldots \left(C_{b_g k^{g-1} - \cdots - b_{g-1}}^2 + S_{b_g k^{g-1} - \cdots - b_{g-1}}^2 \right)^{\frac{1}{2g}}.$$

In view of Lemma 9.4.1, to find the LSD of $a_n^{-1} A_{k,n}$ where $n = k^g + 1$, it suffices to consider the ESD of $\{ \lambda_{(b_1, b_2, \ldots, b_g)} : (b_1, \ldots, b_g) \in T_n \}$: if these have an LSD F, then the LSD of $a_n^{-1} A_{k,n}$ will be (r, θ) in polar coordinates where r is distributed according to F, and θ is distributed uniformly across all the $(2g)$-th roots of unity, and r and θ are independent. With this in mind, define

$$L_{A_{k,n}}(A, \omega) = \frac{1}{\#T_n} \sum_{(b_1, \ldots, b_g) \in T_n} \mathbb{I}(\lambda_{(b_1, \ldots, b_g)} \in A),$$

$$\tilde{L}_{A_{k,n}}(A, \omega) = \frac{1}{\#T_n} \sum_{(b_1, \ldots, b_g) \in T_n} \mathbb{I}(\tilde{\lambda}_{(b_1, \ldots, b_g)} \in A),$$

where

$$\left\{ \tilde{\lambda}_{(b_1,b_2,\ldots,b_g)}, \tilde{\lambda}_{(b_1,b_2,\ldots,b_g)}\omega_{2g}, \ldots, \tilde{\lambda}_{(b_1,b_2,\ldots,b_g)}\omega_{2g}^{2g-1} : (b_1, b_2, \ldots, b_g) \in T_n \right\}$$

are the eigenvalues of $a_n^{-1} A_{k,n}$ corresponding to the major partition sets with input sequence $\{X_i - c_n\}$, $c_n = \mathbb{E}[X_1 \mathbb{I}(|X_1| \le a_n)]$. Further, let $\{\Gamma_j\}$, $\{Z_j\}$ and $\{B_j\}$ be as defined earlier. Let $\{U_{t,j}^*\}$ be a sequence of i.i.d. $U(0,1)$ random variables. Finally, for $1 \le j \le g$ define

$$\tilde{L}_j(\omega) =$$

$$\Big(\sum_{t=1}^{\infty} \sin(2\pi U_{t,j}^*)(B_t(\omega)Z_t(\omega) - \mu_t) \Big)^2 + \Big(\sum_{t=1}^{\infty} \cos(2\pi U_{t,j}^*)(B_t(\omega)Z_t(\omega) - \mu_t) \Big)^2,$$

$$L_j(\omega) = \Big(\sum_{t=1}^{\infty} \sin(2\pi U_{t,j}^*)B_t(\omega)Z_t(\omega) \Big)^2 + \Big(\sum_{t=1}^{\infty} \cos(2\pi U_{t,j}^*)B_t(\omega)Z_t(\omega) \Big)^2.$$

We now state the following theorem.

Theorem 9.4.2 (Bose et al. (2011a)). Let $n = k^g + 1$.

(a) Then $\tilde{L}_{A_{k,n}} \xrightarrow{D} \tilde{L}_{A_k}$, where $\tilde{L}_{A_k}(\cdot, \omega)$ is the random distribution induced by $\tilde{L}_1(\omega)^{1/2g} \tilde{L}_2(\omega)^{1/2g} \cdots \tilde{L}_g(\omega)^{1/2g}$.

(b) Then $L_{A_{k,n}} \xrightarrow{D} L_{A_k}$, where $L_{A_k}(\cdot, \omega)$ is the random distribution induced by $L_1(\omega)^{1/2g} \cdots L_g(\omega)^{1/2g}$.

We recall the sequence of random variables $\{Y_{nk}\}$, $\{B_{nk}\}$, $\{\pi_{nk}\}$ and $\{U_{nk}\}$ defined in Section 9.2. We order the X_k's, $1 \le k \le n$ by their magnitudes $|X_{n1}| \ge |X_{n2}| \ge \cdots \ge |X_{nn}|$. Let $Y_{nk} = |X_{nk}|$ and B_{nk} be the random signs so that $X_{nk} = B_{nk}Y_{nk}$. Let $\pi_{n1}, \pi_{n2}, \ldots, \pi_{nn}$ be the anti-ranks of $Y_{n1}, Y_{n2}, \ldots, Y_{nn}$, $U_{nk} = \frac{\pi_{nk}}{n}$ and $Z_{nk} = \frac{1}{a_n} Y_{nk}$.

Now we state a key lemma which shall be used in the proof of Theorem 9.4.2 and the theorems in the next section. This lemma is a suitable modification of Lemma 9.2.3.

Lemma 9.4.3. Let $n = k^g + 1$ and $\{\pi_{nj}\}_{\{j=1,\ldots,m\}}$, $n \ge 1$ satisfy (9.3) and

$$\frac{k\pi_{nj}}{n}(\bmod 1) \to v_{1,j} \in (0,1), \ldots, \frac{k^{g-1}\pi_{nj}}{n}(\bmod 1) \to v_{g-1,j} \in (0,1), \quad (9.6)$$

where $\{v_{1,j}, \ldots, v_{g-1,j}\}$ are rationally independent. Let

$$\tilde{\pi}_{nj}(b_1, b_2, \ldots, b_g)$$
$$= \Big(\frac{\pi_{nj}}{n}(b_1 k^{g-1} + \cdots + b_g)(\bmod 1), \ldots, \frac{\pi_{nj}}{n}(b_g k^{g-1} - \cdots - b_{g-1})(\bmod 1) \Big),$$

and $\tilde{R}_n$ be a random vector which takes values from the set

$$(\tilde{\pi}_{n1}(b_1, b_2, \ldots, b_g), \ldots, \tilde{\pi}_{nm}(b_1, b_2, \ldots, b_g))$$

with probability $1/\#T_n$ for each $(b_1, b_2, \ldots, b_g) \in T_n$. Then $\tilde{\boldsymbol{R}}_n$ converges weakly to the uniform distribution on the unit cube $[0, 1]^{gm}$.

Proof. We will give a proof only for the case $g = 2$. The proof for the general case is along similar lines. It is enough to show that

$$\int \exp(2\pi i \boldsymbol{w}.\boldsymbol{x}) d\tilde{\boldsymbol{R}}_n(\boldsymbol{x}) \to 0, \tag{9.7}$$

where $\boldsymbol{w} = (w_1, w_2, \ldots, w_{2m})$ is any vector in $\mathbb{R}^{2m}$ with integer coordinates. Note that we can write the left side of (9.7) as

$$\frac{1}{\#T_n} \sum_{(a,b) \in T_n} \exp\left(2\pi i(ak + b) \sum_{j=1}^{m} w_{2j-1} \frac{\pi nj}{n} + 2\pi i(bk - a) \sum_{j=1}^{m} w_{2j} \frac{\pi nj}{n}\right).$$

As $\#T_n \sim n$, we can write the above expression as

$$\approx \frac{1}{n} \sum_{1 \le a \le k} \sum_{1 \le b \le k} \exp(2\pi i a x_{kn}) \exp(2\pi i b y_{kn}), \tag{9.8}$$

where

$$x_{kn} = \sum_{j=1}^{m} w_{2j-1} k \frac{\pi nj}{n} - \sum_{j=1}^{m} w_{2j} \frac{\pi nj}{n} \quad \text{and}$$

$$y_{kn} = \sum_{j=1}^{m} w_{2j} k \frac{\pi nj}{n} + \sum_{j=1}^{m} w_{2j-1} \frac{\pi nj}{n}.$$

Note that both x_{kn} and y_{kn} can be considered with (mod 1), and

$$x_{kn}(\text{mod } 1) \to \sum_{j=1}^{m} w_{2j-1} v_j - \sum_{j=1}^{m} w_{2j} u_j =: x^*.$$

By rational independence of $\{u_j, v_j\}$, for non-zero integers $\{w_j\}$ we have $x^* \neq 0 \pmod 1$. Hence, for all large n, x_{kn} is not an integer. Also note that for all large n, this is bounded away from zero and one. Similar conclusions hold for $\{y_{kn}\}$. Hence, summing the above geometric series (9.8) we get

$$\frac{1}{n} \exp\left(2\pi i x_{kn}\right) \frac{1 - \exp\left(2\pi i k x_{kn}\right)}{1 - \exp\left(2\pi i x_{kn}\right)} \exp\left(2\pi i y_{kn}\right) \frac{1 - \exp\left(2\pi i k y_{kn}\right)}{1 - \exp\left(2\pi i y_{kn}\right)}.$$

Now using the fact that the numerator is bounded, and $\{x_{kn}\}$ and $\{y_{kn}\}$ stay bounded away from 0 and 1 for sufficiently large n, it easily follows that the above expression goes to zero. This completes the proof. $\square$

9.4.1 Proof of Theorem 9.4.2

We shall prove the theorem only for $g = 2$. For $g > 2$, the argument will be similar but with more complicated algebraic calculations. Now define

$$h(x,y) = (\cos(2\pi x), \sin(2\pi x), \cos(2\pi y), \sin(2\pi y)), \quad \text{for } (x,y) \in \mathbb{R},$$

$$\tilde{X}_n(a,b) = \frac{1}{a_n}\left(\tilde{C}_{ak+b}, \tilde{S}_{ak+b}, \tilde{C}_{bk-a}, \tilde{S}_{bk-a}\right), \quad \text{for } (a,b) \in T_n,$$

where

$$\tilde{C}_t = \sum_{j=1}^{n}(X_j - c_n)\cos(\frac{2\pi jt}{n}) \quad \text{and} \quad \tilde{S}_t = \sum_{j=1}^{n}(X_j - c_n)\sin(\frac{2\pi jt}{n}).$$

Let $\tilde{P}_n(\cdot, \omega)$ be the empirical distribution function of $\tilde{X}_n(a,b)$ defined by,

$$\frac{1}{\#T_n}\sum_{(a,b)\in T_n}\mathbb{I}(\tilde{X}_n(a,b) \in \cdot).$$

If we can show that $\tilde{P}_n$ converges in distribution then the result will follow as the empirical measure of $f(\tilde{X}_n(a,b))$ will converge in distribution, where

$$f(x_1, x_2, x_3, x_4) = (x_1^2 + x_2^2)^{\frac{1}{4}}(x_3^2 + x_4^2)^{\frac{1}{4}}.$$

As discussed, we ignore the eigenvalues coming from the partition sets $\mathcal{P}_j$ with $\#\mathcal{P}_j < 4$. Now

$$\tilde{X}_n(a,b) = \sum_{j=1}^{n}h\left(\frac{(ak+b)j}{n}, \frac{(bk-a)j}{n}\right)\frac{X_j - \mu_n}{a_n}$$

$$= \sum_{j=1}^{n}h\big((ak+b)U_{nj}, (bk-a)U_{nj}\big)\big(B_{nj}Z_{nj} - \mathrm{E}\left[B_{nj}Z_{nj}\mathbb{I}(Z_{nj} \leq 1)\right]\big).$$

Let

$$W_{nj}(a,b) = \big((ak+b)U_{nj}(\mathrm{mod}\ 1), (bk-a)U_{nj}(\mathrm{mod}\ 1)\big).$$

To study the behavior of $\tilde{P}_n$, we need to choose $(W_{n1}^*(a,b), \ldots, W_{nn}^*(a,b), \mathbf{0}, \ldots)$ at random from the set of sequences

$$\{(W_{n1}(a,b), \ldots, W_{nn}(a,b), \mathbf{0}, \ldots), (a,b) \in T_n\}.$$

Let

$$W_{nj}^* = (U_{nj}^*, V_{nj}^*).$$

So $\tilde{P}_n$ produces the random variable

$$Y_n^*(\omega) = \sum_{j=1}^{\infty}h(U_{nj}^*, V_{nj}^*)(B_{nj}(\omega)Z_{nj}(\omega) - \mathrm{E}\left[B_{nj}Z_{nj}\mathbb{I}(Z_{nj} \leq 1)\right]).$$

To derive the convergence of $W_{nj}^* = (U_{nj}^*, V_{nj}^*)$, we apply Lemma 9.4.3. The following lemma ensures that the anti-ranks satisfy the conditions of Lemma 9.4.3.

Lemma 9.4.4. Let π_{nj} be the anti-ranks as defined in the previous section. Then $k^s \frac{\pi_{nj}}{n} (\text{mod } 1)$ converges in distribution to the uniform distribution on $(0,1)$, where $s = 0, 1, 2, \ldots, g-1$. Further,

$$\left(\frac{\pi_{n1}}{n}, \ldots, k^{g-1}\frac{\pi_{n1}}{n}(\text{mod } 1), \ldots, \frac{\pi_{nn}}{n}, \ldots, k^{g-1}\frac{\pi_{nn}}{n}(\text{mod } 1) \right)$$
$$\xrightarrow{\mathcal{D}} (\tilde{U}_{1,1}, \ldots, \tilde{U}_{g,1}, \tilde{U}_{1,2}, \ldots, \tilde{U}_{g,2}, \cdots).$$

Proof. The proof goes along the same lines as the proof of Lemma 9.4.3 and easily follows when one considers equation (9.7). We skip the details. □

Proof of Theorem 9.4.2. The main idea of the proof is similar to the idea of the proof of Theorem 5 of Knight (1991) as discussed in Section 9.2. We indicate the main steps. Let

$$Y^*(\omega) = \sum_{j=1}^{\infty} h(U_j^*, V_j^*)(B_j(\omega)Z_j(\omega) - \mathrm{E}\,[B_j Z_j \mathbb{I}(Z_j \leq 1)]).$$

We show that $Y_n^* \xrightarrow{\mathcal{D}} Y^*$.

Let $\tilde{P}$ and $\tilde{E}$ be the **probability** and **expectation** induced by the randomness of $\tilde{P}_n(\cdot, \omega)$ (or equivalently induced by W_{nj}^*).

Now note that $Y_n^*(\omega)$ can be broken into the following two parts:

$$\sum_{j=1}^{\infty} h(U_{nj}^*, V_{nj}^*)(B_{nj}(\omega)Z_{nj}(\omega)\mathbb{I}(Z_{nj}(\omega) > \varepsilon) - \mathrm{E}\,[B_{nj}Z_{nj}\mathbb{I}(\varepsilon < Z_{nj} \leq 1)])$$

$$(9.9)$$

and

$$\sum_{j=1}^{\infty} h(U_{nj}^*, V_{nj}^*)(B_{nj}(\omega)Z_{nj}(\omega)\mathbb{I}(Z_{nj}(\omega) \leq \varepsilon) - \mathrm{E}\,[B_{nj}Z_{nj}\mathbb{I}(Z_{nj} \leq \varepsilon)]). \quad (9.10)$$

We show that the expression in (9.9) converges in distribution almost surely (with respect to the probability measure on Ω) and the expression in (9.10) goes to zero in L_2 in probability (with respect to the probability measure on Ω).

Since h is a bounded function, it follows directly from Lemma 9.2.2 of Knight (1991) that

$$\left\| \sum_{j=1}^{\infty} h(U_{nj}^*, V_{nj}^*)\left[B_{nj}(\omega)Z_{nj}(\omega)\mathbb{I}(Z_{nj} > \varepsilon) - B_j Z_j \mathbb{I}(Z_j > \varepsilon) \right] \right\| \to 0 \quad \text{a.s.,}$$

$$(9.11)$$

and

$$\left\| \sum_{j=1}^{\infty} h(U_{nj}^*, V_{nj}^*)\left[\mathrm{E}\,[B_{nj}Z_{nj}\mathbb{I}(\varepsilon < Z_{nj} \leq 1)] \right. \right.$$
$$\left. \left. - \mathrm{E}\,[B_j(w)Z_j(w)\mathbb{I}(\varepsilon < Z_j \leq 1)] \right] \right\| \to 0. \quad (9.12)$$

Now note that, if $g : \mathbb{R}^{2m} \to \mathbb{R}^4$ is a bounded continuous map having periodicity one in each coordinate, then under the assumptions of Lemma 9.4.3 it follows that

$$\frac{1}{n} \sum_{(a,b) \in T_n} g\left((ak+b)\frac{\pi n1}{n}, (bk-a)\frac{\pi n1}{n}, \ldots, (ak+b)\frac{\pi nm}{n}, (bk-a)\frac{\pi nm}{n}\right)$$

$$\to \int_0^1 \cdots \int_0^1 g(x_1, \ldots, x_{2m}) dx_1 \cdots dx_{2m}.$$

Now we use

$$g(x_1, \ldots, x_{2m}) = \sum_{j=1}^m h(x_{2j-1}, x_{2j})(B_j Z_j \mathbb{I}(Z_j > \varepsilon) - \mathrm{E}\left[B_j Z_j \mathbb{I}(\varepsilon < Z_j \leq 1)\right])$$

and Lemma 9.4.4 to conclude that, for fixed m,

$$\sum_{j=1}^m h(U_{nj}^*, V_{nj}^*)(B_j Z_j \mathbb{I}(Z_j > \varepsilon) - \mathrm{E}\left[B_j Z_j \mathbb{I}(\varepsilon < Z_j \leq 1)\right])$$

$$\xrightarrow{\tilde{D}} \sum_{j=1}^m h(U_j^*, V_j^*)(B_j Z_j \mathbb{I}(Z_j > \varepsilon) - \mathrm{E}\left[B_j Z_j \mathbb{I}(\varepsilon < Z_j \leq 1)\right]),$$

as $n \to \infty$. Here $\tilde{D}$ denotes convergence in distribution with respect to $\tilde{P}$. Since $Z_j \to 0$ as $j \to \infty$ almost surely and $\|h(x,y)\| \leq 2$ for all $(x,y) \in \mathbb{R}^2$, we have

$$\left\| \sum_{k=m+1}^\infty h(U_{nj}^*, V_{nj}^*)(B_j Z_j \mathbb{I}(Z_j > \varepsilon)) \right\| \leq 2 \sum_{j=m+1}^\infty Z_j \mathbb{I}(Z_j > \varepsilon) \to 0, \text{ a.s.,}$$

as $n \to \infty$ and $m \to \infty$. Similarly,

$$\left\| \sum_{j=m+1}^\infty h(U_{nj}^*, V_{nj}^*) \mathrm{E}\left[B_j Z_j \mathbb{I}(\varepsilon < Z_j \leq 1)\right] \right\| \leq 2 \sum_{j=m+1}^\infty \mathrm{E}(Z_j \mathbb{I}(\varepsilon < Z_j \leq 1)) \to 0,$$

as $n \to \infty$ and $m \to \infty$, since $\mathrm{E}(Z_j \mathbb{I}(\varepsilon < Z_j \leq 1)) = O(\varepsilon^{-j}/\Gamma(j))$ as $j \to \infty$. Now

$$\sum_{j=1}^\infty h(U_j^*, V_j^*)(B_j Z_j \mathbb{I}(Z_j > \varepsilon) - \mathrm{E}\left[B_j Z_j \mathbb{I}(\varepsilon < Z_j \leq 1)\right])$$

is finite for almost all ω, with $\tilde{P}$ (defined at the beginning of this proof) probability 1. So, as $m \to \infty$, we have

$$\sum_{j=1}^m h(U_j^*, V_j^*)(B_j Z_j \mathbb{I}(Z_j > \varepsilon) - \mathrm{E}\left[B_j Z_j \mathbb{I}(\varepsilon < Z_j \leq 1)\right])$$

$$\xrightarrow{\tilde{P}} \sum_{j=1}^\infty h(U_j^*, V_j^*)(B_j Z_j \mathbb{I}(Z_j > \varepsilon) - \mathrm{E}\left[B_j Z_j \mathbb{I}(\varepsilon < Z_j \leq 1)\right]).$$

Therefore

$$\sum_{j=1}^{m} h(U_{nj}^*, V_{nj}^*)(B_{nj} Z_{nj} \mathbb{I}(Z_{nj} > \varepsilon) - \mathrm{E}\left[B_{nj} Z_{nj} \mathbb{I}(\varepsilon < Z_{nj} \leq 1)\right])$$

$$\overset{\tilde{\mathcal{D}}}{\to} \sum_{j=1}^{\infty} h(U_j^*, V_j^*)\left(B_j Z_j \mathbb{I}(Z_j > \varepsilon) - \mathrm{E}\left[B_j Z_j \mathbb{I}(\varepsilon < Z_j \leq 1)\right]\right),$$

as $n \to \infty$, in probability. Now to complete the proof part (a), we need to show that

$$\tilde{\mathrm{E}}\left[\left(\sum_{j=1}^{\infty} h(U_{nj}^*, V_{nj}^*)(B_{nj} Z_{nj} \mathbb{I}(Z_{nj} \leq \varepsilon) - \mathrm{E}(B_{nj} Z_{nj} \mathbb{I}(Z_{nj} \leq \varepsilon)))\right)^2\right] \overset{\mathcal{P}}{\to} 0,$$

as $n \to \infty$ and $\varepsilon \to 0$. Now observe that

$$\tilde{\mathrm{E}}\left[\left(\sum_{j=1}^{\infty} h(U_{nj}^*, V_{nj}^*)(B_{nj} Z_{nj} \mathbb{I}(Z_{nj} \leq \varepsilon) - \mathrm{E}(B_{nj} Z_{nj} \mathbb{I}(Z_{nj} \leq \varepsilon)))\right)^2\right]$$

$$= \frac{1}{\#T_n} \sum_{(a,b)\in T_n} \left(\frac{1}{a_n}\sum_{j=1}^{n} h\left(\frac{(ak+b)j}{n}, \frac{(bk-a)j}{n}\right)\right.$$

$$\left.\left(X_j\mathbb{I}(|X_j| \leq a_n\varepsilon) - \mathrm{E}(X_j\mathbb{I}(|X_j| \leq a_n\varepsilon)))\right)^2.$$

Now the expectation with respect to E of the last expression is bounded by

$$2^2 n a_n^{-2} \mathrm{E}\left(X_1^2 \mathbb{I}(|X_1| \leq a_n\varepsilon)\right),$$

and this expression goes to 0 as $n \to \infty$ and $\varepsilon \to 0$. This completes the proof of the first part of the theorem.

The proof of the second part is similar. We indicate the main steps. First, following arguments similar to those given in part (a), as $n \to \infty$,

$$\sum_{j=1}^{\infty} h(U_{nj}^*, V_{nj}^*) B_{nj} Z_{nj} \mathbb{I}(Z_{nj} > \varepsilon) \overset{\tilde{\mathcal{D}}}{\to} \sum_{j=1}^{\infty} h(U_j^*, V_j^*) B_j Z_j \mathbb{I}(Z_j > \varepsilon).$$

Proof of the next step needs extra care. To show that

$$\tilde{P}\left[\|\sum_{j=1}^{\infty} h(U_j^*, V_j^*)\left(B_j Z_j \mathbb{I}(Z_j \leq \varepsilon)\right)\| > \gamma\right] \overset{\mathcal{P}}{\to} 0,$$

as $n \to \infty$ and $\varepsilon \to 0$, observe that

$$\tilde{P}\Big[\|\sum_{j=1}^{\infty} h(U_{nj}^*, V_{nj}^*) B_{nj} Z_{nj} \mathbb{I}(Z_{nj} \le \varepsilon)\| > \gamma\Big]$$

$$\le \tilde{P}\Big[|\sum_{j=1}^{\infty} \cos(2\pi U_{nj}^*) B_{nj} Z_{nj} \mathbb{I}(Z_{nj} \le \varepsilon)| > \frac{\gamma}{2}\Big] + \cdots$$

$$+ \tilde{P}\Big[|\sum_{j=1}^{\infty} \sin(2\pi V_{nj}^*) B_{nj} Z_{nj} \mathbb{I}(Z_{nj} \le \varepsilon)| > \frac{\gamma}{2}\Big]. \qquad (9.13)$$

First observe that

$$\tilde{P}\Big[|\sum_{j=1}^{\infty} \cos(2\pi U_{nj}^*) B_{nj} Z_{nj} \mathbb{I}(Z_{nj} \le \varepsilon)| > \frac{\gamma}{2}\Big] \xrightarrow{P} 0.$$

In fact, we have

$$\tilde{P}\Big[|\sum_{j=1}^{\infty} \cos(2\pi U_{nj}^*) B_{nj} Z_{nj} \mathbb{I}(Z_{nj} \le \varepsilon)| > \frac{\gamma}{2}\Big]$$

$$= \frac{1}{\#T_n} \#\Big\{(a,b) : \big|a_n^{-1} \sum_{l=1}^{n} \cos\big(\frac{2\pi(ak+b)l}{n}\big) X_l \mathbb{I}(|X_l| \le a_n\varepsilon)\big| > \frac{\gamma}{2}\Big\}$$

$$= \frac{1}{\#T_n} \#\Big\{(a,b) \in A_n : \big|a_n^{-1} \sum_{l=1}^{n} \cos\big(\frac{2\pi(ak+b)l}{n}\big) X_l \mathbb{I}(|X_l| \le a_n\varepsilon)\big| > \frac{\gamma}{2}\Big\}$$

$$+ \frac{1}{\#T_n} \#\Big\{(a,b) \in A_n^c : \big|a_n^{-1} \sum_{l=1}^{n} \cos\big(\frac{2\pi(ak+b)l}{n}\big) X_l \mathbb{I}(|X_l| \le a_n\varepsilon)\big| > \frac{\gamma}{2}\Big\},$$

where

$$A_n = \{(a,b) : \sum_{l=1}^{n} \cos(\frac{2\pi(ak+b)l}{n}) \ne 0\}.$$

However,

$$\#A_n = o(n) \text{ and } \frac{\#T_n}{n} \to 1.$$

So

$$\frac{1}{\#T_n}\#\Big\{(a,b):\Big|a_n^{-1}\sum_{l=1}^{n}\cos\Big(\frac{2\pi(ak+b)l}{n}\Big)X_l\mathbb{I}(|X_l|\le a_n\varepsilon)\Big|>\frac{\gamma}{2}\Big\}$$

$$=\frac{o(n)}{n}+\frac{1}{\#T_n}\#\Big\{(a,b)\in A_n^c:$$

$$\Big|a_n^{-1}\sum_{l=1}^{n}\cos\Big(\frac{2\pi(ak+b)l}{n}\Big)X_l\mathbb{I}(|X_l|\le a_n\varepsilon)\Big|>\frac{\gamma}{2}\Big\}$$

$$=\frac{o(n)}{n}+\frac{1}{\#T_n}\#\Big\{(a,b)\in A_n^c:\Big|a_n^{-1}\sum_{l=1}^{n}\cos\Big(\frac{2\pi(ak+b)l}{n}\Big)$$

$$\big[X_l\mathbb{I}(|X_l|\le a_n\varepsilon)-E(X_l\mathbb{I}(|X_l|\le a_n\varepsilon))\big]\Big|>\frac{\gamma}{2}\Big\}$$

$$\le\frac{o(n)}{n}+\frac{1}{\#T_n}\#\Big\{(a,b):\Big|a_n^{-1}\sum_{l=1}^{n}\cos\Big(\frac{2\pi(ak+b)l}{n}\Big)$$

$$\big[X_l\mathbb{I}(|X_l|\le a_n\varepsilon)-E(X_l\mathbb{I}(|X_l|\le a_n\varepsilon))\big]\Big|>\frac{\gamma}{2}\Big\}$$

$$=\frac{o(n)}{n}+\tilde{P}\Big[\Big|\sum_{j=1}^{\infty}\cos(2\pi U_{nj}^*)(B_{nj}Z_{nj}\mathbb{I}(Z_{nj}\le\varepsilon)$$

$$-E(B_{nj}Z_{nj}\mathbb{I}(Z_{nj}\le\varepsilon)))\Big|>\frac{\gamma}{2}\Big]$$

$$\xrightarrow{P}0,$$

by the proof of Theorem 9.4.2(a). Now a similar conclusion holds for the other three terms of (9.13). So $\tilde{P}\Big[\big\|\sum_{j=1}^{\infty}h(U_{nj}^*,V_{nj}^*)\big(B_{nj}Z_{nj}\mathbb{I}(Z_{nj}\le\varepsilon)\big\|>\gamma)\Big]\xrightarrow{P}0$, as $n\to\infty$ and $\varepsilon\to0$. So, this completes the proofs for both parts (a) and (b) of Theorem 9.4.2.   □

9.5   k-circulant: $n=k^g-1$

As before, $n'=n$ and $k'=k$. Now the eigenvalue partition of $\{0,1,2,\ldots,n-1\}$ contains approximately $q=\big[\frac{n}{g}\big]$ sets of size g which are the major partition sets. The remaining sets do not contribute to the LSD. For detailed explanation see the first part of the proof of Theorem 4.5.7. Similar to the developments in the previous section, now the major partition sets $\{S(x)\}$ may be listed as

$$\{b_1k^{g-1}+b_2k^{g-2}+\cdots+b_g,b_2k^{g-1}+\cdots+b_gk+b_1,\ldots,b_gk^{g-1}+b_1k^{g-2}+\cdots+b_{g-1}\},$$

where

$$0\le b_1\le k-1,\ \ldots,0\le b_{g-1}\le k-1,1\le b_g\le k,$$

with $(b_1, b_2, \ldots, b_g) \neq (k-1, k-1, \ldots, k-1)$ and $(b_1, b_2, \ldots, b_g) \neq (k-1, k-1, \ldots, k-1, k)$. Now define

$$T'_n = \{(b_1, b_2, \ldots, b_g) : 0 \leq b_j \leq k-1, \text{ for } 1 \leq j < g \text{ and } 1 \leq b_g \leq k,$$
$$(b_1, b_2, \ldots, b_g) \neq (k-1, k-1, \ldots, k-1) \text{ and}$$
$$(b_1, b_2, \ldots, b_g) \neq (k-1, k-1, \ldots, k-1, k)\},$$

$$\gamma_{(b_1, b_2, \ldots, b_g)} = \frac{1}{a_n} \left(C_{b_1 k^{g-1} + \cdots + b_g} + i S_{b_1 k^{g-1} + \cdots + b_g} \right) \cdots$$
$$\cdots \left(C_{b_g k^{g-1} + \cdots + b_{g-1}} + i S_{b_g k^{g-1} + \cdots + b_{g-1}} \right),$$
$$\eta_{(b_1, b_2, \ldots, b_g)} = |\gamma_{(b_1, b_2, \ldots, b_g)}|^{1/g} \exp\left\{ \frac{i \arg\left(\gamma_{(b_1, b_2, \ldots, b_g)} \right)}{g} \right\}.$$

Then the eigenvalues of the k-circulant $a_n^{-1} A_{k,n}$ with $n = k^g - 1$ corresponding to the partition set $S(b_1 k^{g-1} + b_2 k^{g-2} + \cdots + b_g)$ are

$$\eta_{(b_1, b_2, \ldots, b_g)}, \eta_{(b_1, b_2, \ldots, b_g)} \omega_g, \eta_{(b_1, b_2, \ldots, b_g)} \omega_g^2, \ldots, \eta_{(b_1, b_2, \ldots, b_g)} \omega_g^{g-1},$$

where ω_g is a primitive g-th root of unity. So, to find the LSD, it suffices to consider the ESD of $\{\gamma_{(b_1, b_2, \ldots, b_g)} : (b_1, \ldots, b_g) \in T'_n\}$: if these have an LSD F, then the LSD of $a_n^{-1} A_{k,n}$ will be (r', θ) where r' is distributed according to $h(F)$ where $h(z) = |z|^{1/g} e^{\frac{i \arg(z)}{g}}$ and θ is distributed uniformly across all the g-th roots of unity, and r' and θ are independent. Hence, define

$$L_{A_{k,n}}(A, \omega) = \frac{1}{\#T'_n} \sum_{(b_1, \ldots, b_g) \in T'_n} \mathbb{I}(\gamma_{(b_1, \ldots, b_g)} \in A),$$

$$\tilde{L}_{A_{k,n}}(A, \omega) = \frac{1}{\#T'_n} \sum_{(b_1, \ldots, b_g) \in T'_n} \mathbb{I}(\tilde{\gamma}_{(b_1, \ldots, b_g)} \in A),$$

where $\{\tilde{\gamma}_{(b_1, \ldots, b_g)}\}$ are same as $\{\gamma_{(b_1, \ldots, b_g)}\}$ with the input sequence is $\{X_i - c_n\}$, $c_n = \mathbb{E}[X_1 \mathbb{I}(|X_1| \leq a_n)]$. For $1 \leq j \leq g$, and with $U^*_{t,j}, B_t, Z_t$ as defined in Section 9.4, define

$$\tilde{L}_j(\omega) = \left(\sum_{t=1}^{\infty} \cos(2\pi U^*_{t,j})(B_t(\omega) Z_t(\omega) - \mu_t) \right)$$
$$+ i \left(\sum_{t=1}^{\infty} \sin(2\pi U^*_{t,j})(B_t(\omega) Z_t(\omega) - \mu_t) \right),$$

$$L_j(\omega) = \left(\sum_{t=1}^{\infty} \cos(2\pi U^*_{t,j}) B_t(\omega) Z_t(\omega) \right) + i \left(\sum_{t=1}^{\infty} \sin(2\pi U^*_{t,j}) B_t(\omega) Z_t(\omega) \right).$$

Theorem 9.5.1 (Bose et al. (2011a)). Let $n = k^g - 1$.

(a) Let $\tilde{L}_{A_{k,n}}(\cdot, \omega)$ be the ESD of $\{\tilde{\gamma}_{(b_1, \ldots, b_g)} : (b_1, \ldots, b_g) \in T'_n\}$. Then

$\tilde{L}_{A_{k,n}} \xrightarrow{D} \tilde{L}_{A_k}$, where $\tilde{L}_{A_k}(\cdot, \omega)$ is the random distribution induced by $\tilde{L}_1(\omega)\tilde{L}_2(\omega)\cdots\tilde{L}_g(\omega)$.

(b) Then $L_{A_{k,n}} \xrightarrow{D} L_{A_k}$, where $L_{A_k}(\cdot, \omega)$ is the random distribution induced by $L_1(\omega)L_2(\omega)\ldots L_g(\omega)$.

The proof of Theorem 9.5.1 is similar to the proof of Theorem 9.4.2, so we skip it.

9.6   Tail of the LSD

Now we show that even though the input sequence is heavy-tailed, the LSDs (in the almost sure sense) are light-tailed. Indeed for $\alpha \in (0, 1)$ they have bounded support.

Lemma 9.6.1. (a) For $\alpha \in (0, 1)$, the variable $\sum_{t=1}^{\infty} \cos(2\pi U_t^*)B_t(\omega)Z_t(\omega)$ has bounded support for almost all $\omega \in \Omega$.

(b) For $\alpha \in [1, 2)$, the variable $\sum_{t=1}^{\infty} \cos(2\pi U_t^*)B_t(\omega)Z_t(\omega)$ has light-tail for almost all $\omega \in \Omega$.

It is clear from Lemma 9.6.1(a) that the LSD obtained in Theorems 9.4.2(b), 9.5.1(b) and 9.3.2(b) have bounded support for $\alpha \in (0, 1)$ and for almost all $\omega \in \Omega$. From Lemma 9.6.1(b), it follows that the LSD in Theorems 9.4.2(b), 9.5.1(b) and 9.3.2(b) have light-tail for $\alpha \in [1, 2)$ and for almost all $\omega \in \Omega$. Similar conclusions hold about the LSD in Theorems 9.4.2(a), 9.5.1(a) and 9.3.2(a).

Now we briefly sketch the proof of Lemma 9.6.1.

Proof. Recall that $Z_j = \Gamma_j^{-1/\alpha} = \left(\sum_{t=1}^j E_t\right)^{-1/\alpha}$, where $\{E_i\}$ is a sequence of i.i.d. exponential random variables with mean 1. Hence, by the Law of the Iterated Logarithm (see Section 11.2 of Appendix), for almost all ω and for arbitrary $\varepsilon > 0$ there exist $j_0(\omega)$ so that for $j \geq j_0(\omega)$,

$$\left(\frac{1}{j + (\sqrt{2} + \varepsilon)\sqrt{j \log\log j}}\right)^{\frac{1}{\alpha}} < Z_j(\omega) < \left(\frac{1}{j - (\sqrt{2} + \varepsilon)\sqrt{j \log\log j}}\right)^{\frac{1}{\alpha}}.$$

Hence

$$-\sum_{t=1}^{j_0(\omega)} Z_t(\omega) - \sum_{t=j_0(\omega)}^{\infty} \left(\frac{1}{t - (\sqrt{2} + \varepsilon)\sqrt{t \log\log t}}\right)^{\frac{1}{\alpha}}$$

$$\leq \sum_{t=1}^{\infty} \cos(2\pi U_t^*)B_t(\omega)Z_t(\omega) \leq \sum_{t=1}^{j_0(\omega)} Z_t(\omega) + \sum_{t=j_0(\omega)}^{\infty} \left(\frac{1}{t - (\sqrt{2} + \varepsilon)\sqrt{t \log\log t}}\right)^{\frac{1}{\alpha}}.$$

So $\sum_{t=1}^{\infty} \cos(2\pi U_t^*) B_t(\omega) Z_t(\omega)$ is bounded for almost all ω when $\alpha \in (0, 1)$. For $\alpha \in [1, 2)$, and for almost all ω,

$$\text{Var}\left(\sum_{t=1}^{\infty} \cos(2\pi U_t^*) B_t(\omega) Z_t(\omega)\right) = \sum_{t=1}^{\infty} B_t^2(\omega) Z_t^2(\omega) \text{Var}(\cos(2\pi U_t^*))$$

$$= \text{Var}(\cos(2\pi U_1^*)) \sum_{t=1}^{\infty} B_t^2(\omega) Z_t^2(\omega)$$

$$\leq \text{Var}(\cos(2\pi U_1^*)) \left(\sum_{t=1}^{j_0(\omega)} Z_t^2(\omega)\right)$$

$$+ \sum_{t=j_0(\omega)}^{\infty} \left(\frac{1}{t - (\sqrt{2} + \varepsilon)\sqrt{t \log \log t}}\right)^{\frac{2}{\alpha}}\right) < \infty,$$

and hence $\sum_{t=1}^{\infty} \cos(2\pi U_t^*) B_t(\omega) Z_t(\omega)$ is light-tailed. $\square$

9.7 Exercises

1. Prove Theorem 9.3.1.
2. Prove Theorem 9.3.2.
3. Prove Lemma 9.2.1 using Lemmata 1 and 2 of LePage et al. (1981).
4. Prove Lemma 9.4.4.
5. Check how (9.11) and (9.12) follow from Lemma 2 of Knight (1991).
6. Look up the proof of Theorem 9.2.4 in Knight (1991).
7. Prove Theorem 9.5.1.

10

Heavy-tailed input: spectral radius

In this chapter the focus is on the distributional convergence of the spectral radius and hence of the spectral norm of the circulant, the reverse circulant and the symmetric circulant matrices when the input sequence is heavy-tailed. The approach is to use the already known methods for the study of the maximum of periodograms for heavy-tailed sequences. This method is completely different from the method used to derive the results in Chapter 5 for light-tailed entries.

10.1 Input sequence and scaling

We recall some of the definitions and notations introduced in Section 9.1 of Chapter 9. Let $\{Z_t, \ t \in \mathbb{Z}\}$ be a sequence of i.i.d. random variables with common distribution F which is in the *domain of attraction* of an α-*stable law* with $0 < \alpha < 1$. Thus, there exist $p, q \geq 0$ with $p + q = 1$ and a *slowly varying* function ℓ, such that as $x \to \infty$,

$$\frac{P(Z_1 > x)}{P(|Z_1| > x)} \to p, \quad \frac{P(Z_1 \leq -x)}{P(|Z_1| > x)} \to q \text{ and } P(|Z_1| > x) \approx x^{-\alpha}\ell(x). \quad (10.1)$$

Let $\{\Gamma_j\}$, $\{U_j\}$ and $\{B_j\}$ be three independent sequences defined on the same probability space where $\Gamma_j = \sum_{i=1}^{j} E_i$ and $\{E_i\}$ are i.i.d. exponential with mean 1, U_j are i.i.d. $U(0,1)$, and B_j are i.i.d. with

$$P(B_1 = 1) = p \quad \text{and} \quad P(B_1 = -1) = q, \quad (10.2)$$

where p and q are as defined in (10.1). We recall the class $S_\alpha(\sigma, \beta, \mu)$ from Section 9.1, and also define

$$Y_\alpha = \sum_{j=1}^{\infty} \Gamma_j^{-1/\alpha} \stackrel{\mathcal{D}}{\cong} S_\alpha(C_\alpha^{-\frac{1}{\alpha}}, 1, 0) \text{ where } C_\alpha = \left(\int_0^\infty x^{-\alpha} \sin x dx \right)^{-1}.$$

$$(10.3)$$

For a non-decreasing function f on $\mathbb{R}$, let $f^{\leftarrow}(y) = \inf\{s : f(s) > y\}$. Then the *scaling sequence* $\{b_n\}$ is defined as

$$b_n = \left(\frac{1}{P[|Z_1| > \cdot]} \right)^{\leftarrow} (n) \approx n^{1/\alpha} \ell_0(n) \text{ for some slowly varying function } \ell_0.$$

10.2   Reverse circulant and circulant

Consider the RC_n with the input sequence $\{Z_t\}$. Then from Section 1.3 of Chapter 1, the eigenvalues $\{\lambda_k,\ 0 \le k \le n-1\}$ of $\frac{1}{b_n}RC_n$ are given by:

$$
\begin{cases}
\lambda_0 & = \frac{1}{b_n}\sum_{t=0}^{n-1} Z_t \\[2mm]
\lambda_{n/2} & = \frac{1}{b_n}\sum_{t=0}^{n-1}(-1)^t Z_t, \quad \text{if } n \text{ is even} \\[2mm]
\lambda_k & = -\lambda_{n-k} = \sqrt{I_n(\omega_k)},\ 1 \le k \le \lfloor \frac{n-1}{2} \rfloor,
\end{cases}
\tag{10.4}
$$

where

$$
I_n(\omega_k) = \frac{1}{b_n^2} \Big| \sum_{t=0}^{n-1} Z_t e^{-it\omega_k} \Big|^2 \quad \text{and} \quad \omega_k = \frac{2\pi k}{n}.
$$

The eigenvalues of $\frac{1}{b_n}C_n$ are given by

$$
\lambda_j = \frac{1}{b_n} \sum_{t=1}^{n} Z_t e^{it\omega_j}, \quad 0 \le j \le n-1.
$$

From the eigenvalue structure of C_n and RC_n, it is clear that

$$
\mathrm{sp}(\frac{1}{b_n}C_n) = \mathrm{sp}(\frac{1}{b_n}RC_n)
$$

and therefore they have identical limiting behavior. This behavior is described in the following result.

Theorem 10.2.1 (Bose et al. (2010)). Suppose the input sequence $\{Z_t\}$ satisfies (10.1). Then for $\alpha \in (0,1)$, both $\mathrm{sp}(\frac{1}{b_n}C_n)$ and $\mathrm{sp}(\frac{1}{b_n}RC_n)$ converge in distribution to Y_α, where Y_α is as in (10.3).

Note that $\{|\lambda_k|^2;\ 1 \le k < n/2\}$ is the periodogram of $\{Z_t\}$ at the frequencies $\{\omega_k;\ 1 \le k < n/2\}$. Mikosch et al. (2000) have established the weak convergence of the maximum of the periodogram based on a heavy-tailed sequence for $0 < \alpha < 1$. The main idea for the proof of Theorem 10.2.1 is taken from their work.

Let $\epsilon_x(\cdot)$ denote the point measure which gives unit mass to any set containing x and let $E = [0,1] \times ([-\infty,\infty]\setminus\{0\})$. Let $M_p(E)$ be the set of point measures on E, topologized by vague convergence. The following convergence result follows from Proposition 3.21 of Resnick (1987):

$$
N_n := \sum_{k=1}^{n} \epsilon_{(k/n, Z_k/b_n)} \overset{\mathcal{V}}{\to} N := \sum_{j=1}^{\infty} \epsilon_{(U_j, B_j \Gamma_j^{-1/\alpha})} \quad \text{in } M_p(E).
\tag{10.5}
$$

Suppose f is a bounded continuous complex valued function defined on $\mathbb{R}$ and without loss of generality assume $|f(x)| \leq 1$ for all $x \in \mathbb{R}$. Now pick $\eta > 0$, and define $T_\eta : M_p(E) \longrightarrow C[0, \infty)$ as follows:

$$(T_\eta m)(x) = \sum_j v_j \mathbb{I}_{\{|v_j| > \eta\}} f(2\pi x t_j),$$

if $m = \sum_j \epsilon_{(t_j, v_j)} \in M_p(E)$ and v_j's are finite. Elsewhere, set $(T_\eta m)(x) = 0$.

Lemma 10.2.2. The map $T_\eta : M_p(E) \longrightarrow C[0, \infty)$ is continuous almost surely with respect to the distribution of N.

Proof. This Lemma was proved by Mikosch et al. (2000) for the function $f(x) = \exp(-ix)$. The same proof works in our case. For the sake of completeness we give the details.

It is enough to show that if $x_n \to x \geq 0$ and $m_n \overset{v}{\to} m$ in $M_p(E)$, where

$$m\{\partial([0, 1] \times \{|v| \geq \eta\}) \cap [0, 1] \times \{-\infty, \infty\}\} = 0,$$

then $(T_\eta m_n)(x_n) \to (T_\eta m)(x)$.

To show this, let

$$m_n(\cdot) = \sum_j \epsilon_{\left(t_j^{(n)}, v_j^{(n)}\right)}(\cdot) \quad \text{and} \quad m(\cdot) = \sum_j \epsilon_{(t_j, v_j)}(\cdot).$$

Consider the set

$$K_\eta := [0, 1] \times \{v : |v| \geq \eta\}.$$

Clearly, K_η is compact in E with $m(\partial K_\eta) = 0$. Since $m_n \overset{v}{\to} m$, we can find an n_0 such that for $n \geq n_0$,

$$m_n(K_\eta) = m(K_\eta) =: l, \text{ say,}$$

and there is an enumeration of the points in K_η such that

$$\left((t_k^{(n)}, v_k^{(n)}), \ 1 \leq k \leq l\right) \to \left((t_k, v_k), \ 1 \leq k \leq l\right).$$

Without loss of generality we can assume that for given $\xi > 0$,

$$\sup_{n \geq n_0} |x_n| \vee \sup_{1 \leq k \leq l} |v_k^{(n)}| \leq \xi.$$

Therefore

$$|(T_\eta m_n)(x_n) - (T_\eta m)(x)| = \left| \sum_{k=1}^{l} v_k^{(n)} f(-2\pi x_n t_k^{(n)}) - \sum_{k=1}^{l} v_k f(-2\pi x t_k) \right|$$

$$\leq \sum_{k=1}^{l} \left| v_k^{(n)} f(-2\pi x_n t_k^{(n)}) - v_k f(-2\pi x t_k) \right|$$

$$\leq \sum_{k=1}^{l} \left| v_k^{(n)} - v_k \right|$$

$$+ \sum_{k=1}^{l} |v_k| \left| f(-2\pi x_n t_k^{(n)}) - f(-2\pi x t_k) \right|,$$

so that

$$\lim_{n \to \infty} |(T_\eta m_n)(x_n) - (T_\eta m)(x)| = 0.$$

This completes the proof of the Lemma. □

Lemma 10.2.3. For $0 < \alpha < 1$ and $0 \le x < \infty$, the following convergence holds in $C[0, \infty)$ as $n \to \infty$:

$$J_{n,Z}(x/n) := \sum_{j=1}^{n} \frac{Z_j}{b_n} f(2\pi x j/n) \xrightarrow{D} J_\infty(x) := \sum_{j=1}^{\infty} B_j \Gamma_j^{-1/\alpha} f(2\pi x U_j).$$

Proof. The proof is similar to the proof of Proposition 2.2 of Mikosch et al. (2000). We briefly sketch the proof in our case.

Applying Lemma 10.2.2 on (10.5) we have

$$J_{n,Z}^{(\eta)}(x/n) := \sum_{j=1}^{n} \frac{Z_j}{b_n} f(2\pi x j/n) \mathbb{I}_{\{|Z_j| > \eta b_n\}}$$

$$\xrightarrow{D} \sum_{j=1}^{\infty} B_j \Gamma_j^{-1/\alpha} f(2\pi x U_j) \mathbb{I}_{\{\Gamma_j^{-1/\alpha} > \eta\}} =: J_\infty^{(\eta)}(x) \quad \text{in} \quad C[0, \infty).$$

Also, as $\eta \to 0$, by DCT we have

$$J_\infty^{(\eta)}(x) \xrightarrow{D} J_\infty(x) := \sum_{j=1}^{\infty} B_j \Gamma_j^{-1/\alpha} f(2\pi x U_j).$$

So using Theorem 3.2 of Billingsley (1995), the proof will be complete if for any $\epsilon > 0$,

$$\lim_{\eta \to 0} \limsup_{n \to \infty} P\left(\|J_{n,Z}^{(\eta)} - J_{n,Z}\|_\infty > \epsilon \right) = 0. \tag{10.6}$$

Here $\|x(\cdot) - y(\cdot)\|_\infty$, the metric distance in $C[0, \infty)$, is given by

$$\|x(\cdot) - y(\cdot)\|_\infty = \sum_{n=1}^{\infty} \frac{1}{2^n} [\|x(\cdot) - y(\cdot)\|_n \wedge 1], \quad \text{where}$$

$$\|x(\cdot) - y(\cdot)\|_n = \sup_{t \in [0,n]} |x(t) - y(t)|.$$

Now

$$\lim_{\eta \to 0} \limsup_{n \to \infty} P\left(\|J_{n,Z}^{(\eta)} - J_{n,Z}\|_\infty > \epsilon \right) \le \lim_{\eta \to 0} \limsup_{n \to \infty} P\left(\sum_{j=1}^{n} |\frac{Z_j}{b_n}| \mathbb{I}_{\{|Z_j| \le \eta b_n\}} > \epsilon \right)$$

$$\le \lim_{\eta \to 0} \limsup_{n \to \infty} n\epsilon^{-1} E\left(|\frac{Z_1}{b_n}| \mathbb{I}_{\{|Z_1| \le \eta b_n\}} \right).$$

By an application of Karamata's theorem (see Resnick (1987), Exercise 0.4.2.8), we get that

$$n E\left(|\frac{Z_1}{b_n}| \mathbb{I}_{\{|Z_1| \le \eta b_n\}} \right) \approx \frac{\alpha}{1 - \alpha} n\eta \, P(|Z_1| > \eta b_n) \approx \frac{\alpha}{1 - \alpha} \eta^{1-\alpha},$$

and $\frac{\alpha}{1-\alpha}\eta^{1-\alpha} \to 0$ as $\eta \to 0$. This completes the proof of the lemma. $\qquad \square$

Proof of Theorem 10.2.1. We use Lemma 10.2.2 and Lemma 10.2.3 with $f(x) = \exp(-ix)$. It is immediate that

$$\frac{1}{b_n}\mathrm{sp}(C_n) \le \frac{1}{b_n}\sum_{t=1}^{n}|Z_t|. \tag{10.7}$$

It is well known that (see Feller (1971))

$$\frac{1}{b_n}\sum_{t=1}^{n}|Z_t| \xrightarrow{\mathcal{D}} Y_\alpha = \sum_{j=1}^{\infty}\Gamma_j^{-1/\alpha} \overset{\mathcal{D}}{\simeq} S_\alpha(C_\alpha^{-1/\alpha},1,0). \tag{10.8}$$

Hence it remains to show that for $\gamma > 0$,

$$\liminf_{n\to\infty} \mathrm{P}\left(\frac{1}{b_n}\mathrm{sp}(C_n) > \gamma\right) \ge \mathrm{P}(Y_\alpha > \gamma). \tag{10.9}$$

Now observe that for any integer K and sufficiently large n,

$$\mathrm{P}\left(\sup_{j=1,\ldots,\lfloor\frac{n}{2}\rfloor}|J_{n,Z}(j/n)| > \gamma\right) \ge \mathrm{P}\left(\sup_{j=1,\ldots,K}|J_{n,Z}(j/n)| > \gamma\right).$$

On the other hand, from Lemma 10.2.3 we have

$$\left(J_{n,Z}(j/n), 1 \le j \le K\right) \xrightarrow{\mathcal{D}} \left(J_\infty(j), 1 \le j \le K\right) \text{ in } \mathbb{R}^k.$$

Hence

$$\sup_{j=1,\ldots,K}|J_{n,Z}(j/n)| \xrightarrow{\mathcal{D}} \sup_{j=1,\ldots,K}|J_\infty(j)|,$$

and so letting $K \to \infty$,

$$\liminf_{n\to\infty}\mathrm{P}\left(\sup_{j=1,\ldots,\lfloor\frac{n}{2}\rfloor}|J_{n,Z}(j/n)| > \gamma\right) \ge \mathrm{P}\left(\sup_{j=1,\ldots,\infty}|J_\infty(j)| > \gamma\right).$$

Now the theorem follows from Lemma 10.2.4 given below. $\qquad \square$

Lemma 10.2.4.

$$\sup_{j=1,\ldots,\infty}|J_\infty(j)| = \sup_{j=1,\ldots,\infty}\left|\sum_{t=1}^{\infty}B_t\Gamma_t^{-1/\alpha}\exp(-2\pi ijU_t)\right| = Y_\alpha \text{ a.s.}$$

Proof. Define

$$\Omega_0 = \left\{\omega \in \Omega : \sum_{j=1}^{\infty}\Gamma_j^{-1/\alpha}(\omega) < \infty \text{ and for all } m \ge 1,\right.$$

$$\left.(U_1(\omega),\ldots,U_m(\omega)) \text{ are rationally independent}\right\}.$$

Then $P(\Omega_0) = 1$. Let $\bar{x}$ denote the fractional part of x. Now by Weyl (1916), for any $\omega \in \Omega_0$,

$$\left(\overline{nU_1(\omega)}, \ldots, \overline{nU_m(\omega)}\right), \ n \in \mathbb{N}$$

is dense in $[0,1]^m$. Fix any $\omega \in \Omega_0$ and $\epsilon > 0$. Then there exist an $N \in \mathbb{N}$ such that $\sum_{j=N+1}^{\infty} \Gamma_j^{-1/\alpha}(\omega) < \epsilon$, and from the results of Weyl (1916) there exist an $N_0 \in \mathbb{N}$ such that

$$\mathcal{R}\left(B_j \exp(-2\pi i N_0 U_j)\right) \geq 1 - \frac{\epsilon}{N\Gamma_j^{-1/\alpha}}, \quad j = 1, \ldots, N.$$

Then we have

$$\sup_{1 \leq j \leq \infty} \left|\sum_{t=1}^{\infty} B_t \Gamma_t^{-1/\alpha} \exp(-2\pi i j U_t)\right| \geq \sup_{1 \leq j \leq \infty} \left|\sum_{t=1}^{N} B_t \Gamma_t^{-1/\alpha} \exp(-2\pi i j U_t)\right|$$

$$- \sum_{t=N+1}^{\infty} \Gamma_t^{-1/\alpha}$$

$$\geq \left|\sum_{t=1}^{N} B_t \Gamma_t^{-1/\alpha} \exp(-2\pi i N_0 U_t)\right| - \epsilon$$

$$\geq \mathcal{R}\left(\sum_{t=1}^{N} B_t \Gamma_t^{-1/\alpha} \exp(-2\pi i N_0 U_t)\right) - \epsilon$$

$$\geq \sum_{t=1}^{N} \left(1 - \frac{\epsilon}{N\Gamma_t^{-1/\alpha}}\right)\Gamma_t^{-1/\alpha} - \epsilon$$

$$= \sum_{t=1}^{N} \Gamma_t^{-1/\alpha} - 2\epsilon.$$

Letting first $N \to \infty$, and then $\epsilon \to 0$, we get $\sup_{j=1,\ldots,\infty} |J_\infty(j)| \geq Y_\alpha$. Trivially $\sup_{j=1,\ldots,\infty} |J_\infty(j)| \leq Y_\alpha$. This completes the proof. □

10.3   Symmetric circulant

We recall that the eigenvalues $\{\lambda_k, \ 0 \leq k \leq n-1\}$ of $\frac{1}{b_n}SC_n$, when the input sequence is $\{Z_t\}$, is given by (see Section 1.2 of Chapter 1):

(i) for n odd:

$$\begin{cases} \lambda_0 & = \frac{1}{b_n}\left[Z_0 + 2\sum_{j=1}^{\lfloor\frac{n}{2}\rfloor} Z_j\right] \\[2mm] \lambda_k & = \frac{1}{b_n}\left[Z_0 + 2\sum_{j=1}^{\lfloor\frac{n}{2}\rfloor} Z_j \cos(\omega_k j)\right], \quad 1 \leq k \leq \lfloor\frac{n}{2}\rfloor, \end{cases}$$

(ii) for n even:

$$
\begin{cases}
\lambda_0 &= \frac{1}{b_n}\left[Z_0 + 2\sum_{j=1}^{\frac{n}{2}-1} Z_j + Z_{n/2}\right] \\[2mm]
\lambda_k &= \frac{1}{b_n}\left[Z_0 + 2\sum_{j=1}^{\frac{n}{2}-1} Z_j \cos(\omega_k j) + (-1)^k Z_{n/2}\right], \quad 1 \le k \le \frac{n}{2},
\end{cases}
$$

with $\lambda_{n-k} = \lambda_k$ in both cases.

Theorem 10.3.1 (Bose et al. (2010)). Assume that the input sequence is i.i.d. $\{Z_t\}$ and satisfies (10.1). Then for $\alpha \in (0,1)$, $\mathrm{sp}(\frac{1}{b_n}SC_n) \overset{D}{\to} 2^{1-1/\alpha}Y_\alpha$, where Y_α is as in (10.3).

Proof. The proof is similar to the proof of Theorem 10.2.1. We provide a sketch of the proof for odd n. For even n, the changes needed are minor. Define

$$
J_{n,Z}(x) := 2\frac{1}{b_n}\sum_{t=1}^{q} Z_t \cos(2\pi x t) \quad \text{and} \quad M_{n,Z} := \max_{0 \le k \le q} \left| J_{n,Z}(k/n) \right|, \quad (10.10)
$$

where $q = q(n) = \lfloor \frac{n}{2} \rfloor$. Since $\left| \mathrm{sp}(\frac{1}{b_n}SC_n) - M_{n,Z} \right| \to 0$ almost surely, it is enough to show that $M_{n,Z} \overset{D}{\to} 2^{1-1/\alpha}Y_\alpha$.

Note that (10.5) holds with $[0,1]$ replaced by $[0,1/2]$, and upon letting $N_n = \sum_{k=1}^{q} \epsilon_{(k/n, Z_k/b_q)}$, $N = \sum_{j=1}^{\infty} \epsilon_{(U_j, B_j \Gamma_j^{-1/\alpha})}$, and U_j to be i.i.d. $U[0, 1/2]$. Now following the arguments given in the proof of Lemma 10.2.2, Lemma 10.2.3, and with $f(x) = \cos x$,

$$
J_{n,Z}(x/n) = 2\frac{1}{b_n}\sum_{k=1}^{q} Z_k \cos\frac{2\pi k x}{n} \overset{D}{\to} 2^{1-1/\alpha}\sum_{j=1}^{\infty} B_j \Gamma_j^{-1/\alpha} \cos(2\pi x U_j).
$$

$$(10.11)$$

It is obvious that

$$
M_{n,Z} \le 2\frac{1}{b_n}\sum_{t=1}^{q}|Z_t| \overset{D}{\to} 2^{1-1/\alpha}\sum_{j=1}^{\infty}\Gamma_j^{-1/\alpha} = 2^{1-1/\alpha}Y_\alpha.
$$

Following the arguments given to prove (10.9), we can show that

$$
\liminf_{n\to\infty} P(M_{n,Z} > \eta) \ge P(2^{1-1/\alpha}Y_\alpha > \eta), \quad \text{for } \eta > 0.
$$

This completes the proof of the theorem. $\qquad\square$

Remark 10.3.1. (i) Theorems 10.2.1 and 10.3.1 are rather easy to derive when $p = 1$, that is, when the left tail is negligible compared to the right tail. Let us consider $\mathrm{sp}(\frac{1}{b_n}RC_n)$. From its eigenvalue structure,

$$
\mathrm{sp}\left(\frac{1}{b_n}RC_n\right) \le \frac{1}{b_n}\sum_{t=1}^{n}|Z_t|.
$$

For the lower bound note that

$$P\left(\text{sp}(\frac{1}{b_n}RC_n) > x\right) \geq P(\lambda_0 > x) = P\left(\frac{1}{b_n}\sum_{t=1}^n Z_t > x\right).$$

Now since $P(|Z_1| > x) \approx P(Z_1 > x)$ as $x \to \infty$, the upper and the lower bounds converge with the same scaling constant and hence Theorem 10.2.1 holds. The details on these convergences can be found in Chapter 1 of Samorodnitsky and Taqqu (1994). A similar conclusion can be drawn for the symmetric circulant matrices too when $p = 1$.

(ii) When the input sequence $\{Z_i\}$ is i.i.d. non-negative and satisfies (10.1) with $\alpha \in (1, 2)$ then from the above arguments it is easy to derive the distributional behavior of the spectral radius. In particular, if

$$k_j = \frac{\alpha}{\alpha-1}\left(j^{\frac{\alpha-1}{\alpha}} - (j-1)^{\frac{\alpha-1}{\alpha}}\right) \quad \text{and} \quad \widetilde{Y}_\alpha = \sum_{j=1}^\infty (\Gamma_j - k_j) \overset{\mathcal{D}}{\approx} S_\alpha(C_\alpha^{-\frac{1}{\alpha}}, 1, 0)$$

then, as $n \to \infty$,

$$P\left(\frac{\text{sp}(RC_n) - n\,\text{E}[Z_1]}{b_n} > x\right) \to P\left(\widetilde{Y}_\alpha > x\right)$$

and

$$P\left(\frac{\text{sp}(SC_n) - n\,\text{E}[Z_1]}{b_n} > x\right) \to P\left(2^{1-1/\alpha}\widetilde{Y}_\alpha > x\right).$$

If $\alpha = 1$ and $\{Z_i\}$ are non-negative, then

$$P\left(\frac{\text{sp}(RC_n) - nb_n\int_0^\infty \sin(\frac{x}{b_n})\,P(Z_1 \in dx)}{b_n} > x\right) \to P\left(\widetilde{\widetilde{Y}}_\alpha > x\right),$$

where $\widetilde{\widetilde{Y}}_\alpha$ is an $S_1(2/\pi, 1, 0)$ random variable. Similar results hold for the symmetric circulant matrix also.

10.4   Heavy-tailed: dependent input

Now suppose that the input sequence is a *linear process* $\{X_t, t \in \mathbb{Z}\}$ and it is given by

$$X_t = \sum_{j=-\infty}^\infty a_j Z_{t-j}, \quad t \in \mathbb{Z}, \quad \text{where} \quad \sum_{j=-\infty}^\infty |a_j|^{\alpha-\varepsilon} < \infty \text{ for some } 0 < \varepsilon < \alpha.$$

$$(10.12)$$

Suppose that $\{Z_i\}$ are i.i.d. random variables and satisfy (10.1) with $0 < \alpha < 1$. Using $E|Z|^{\alpha-\varepsilon} < \infty$ and the assumption on the $\{a_j\}$, we have

$$E|X_t|^{\alpha-\varepsilon} \leq \sum_{j=-\infty}^{\infty} |a_j|^{\alpha-\varepsilon} \, E|Z_{t-j}|^{\alpha-\varepsilon} = E|Z_1|^{\alpha-\varepsilon} \sum_{j=-\infty}^{\infty} |a_j|^{\alpha-\varepsilon} < \infty.$$

Hence X_t is finite almost surely. Let

$$\psi(x) = \sum_{j=-\infty}^{\infty} a_j \exp(-i2\pi xj), \quad x \in [0,1],$$

be the *transfer function* of the linear filter $\{a_j\}$, and

$$f_X(x) = |\psi(x)|^2,$$

be the *power transfer function* of $\{X_t\}$. Define

$$M(RC_n, f_X) = \max_{0 \leq k < \frac{n}{2}} \frac{|\lambda_k|}{\sqrt{f_X(k/n)}}, \quad M(C_n, f_X) = \max_{0 \leq k < \frac{n}{2}} \frac{|\lambda_k|}{\sqrt{f_X(k/n)}},$$

$$M(SC_n, f_X) = \max_{0 \leq k < \frac{n}{2}} \frac{|\lambda_k|}{\sqrt{f_X(k/n)}},$$

where in each case $\{\lambda_k\}$ are the eigenvalues of the corresponding matrix. From the eigenvalue structure of C_n and RC_n, $M(C_n, f_X) = M(RC_n, f_X)$.

Theorem 10.4.1 (Bose et al. (2010)). Suppose that $\{Z_t\}$ is i.i.d and satisfies (10.1). Further, the input sequence is $\{X_n\}$, and $\{a_j\}$ satisfies (10.12). Suppose f_X is strictly positive on $[0, 1/2]$. Then

(a) $M(\frac{1}{b_n}C_n, f_X) \xrightarrow{D} Y_\alpha$ and $M(\frac{1}{b_n}RC_n, f_X) \xrightarrow{D} Y_\alpha$.

(b) Further, if $a_j = a_{-j}$, then $M(\frac{1}{b_n}SC_n, f_X) \xrightarrow{D} 2^{1-1/\alpha}Y_\alpha$.

Proof. (a) The proof is along the lines of the proof of Lemma 2.6 in Mikosch et al. (2000). Let $\widehat{C_n}$ be the circulant matrix formed with independent entries $\{Z_i\}$. To prove the result it is enough to show that

$$\left| M(\frac{1}{b_n}C_n, f_X) - \mathrm{sp}(\frac{1}{b_n}\widehat{C_n}) \right| \xrightarrow{P} 0.$$

Let $J_{n,Z}(x) = \frac{1}{b_n}\sum_{t=1}^{n} Z_t \exp(-i2\pi xt)$. Note that

$$\left| M(\frac{1}{b_n}C_n, f_X) - \mathrm{sp}(\frac{1}{b_n}\widehat{C_n}) \right| = \Big| \sup_{1 \leq k \leq n} (f_X(k/n))^{-1/2}|J_{n,X}(k/n)|$$

$$- \sup_{1 \leq k \leq n} |J_{n,Z}(k/n)| \Big|$$

$$\leq \sup_{1 \leq k \leq n} \left| |\psi(k/n)^{-1}J_{n,X}(k/n)| - |J_{n,Z}(k/n)| \right|$$

$$\leq \sup_{1 \leq k \leq n} \left| \psi(k/n)^{-1}J_{n,X}(k/n) - J_{n,Z}(k/n) \right|,$$

and

$$J_{n,X}(x) = \frac{1}{b_n} \sum_{t=1}^{n} X_t \exp(-i2\pi xt)$$

$$= \frac{1}{b_n} \sum_{j=-\infty}^{\infty} a_j \exp(-i2\pi xj) \left(\sum_{t=1}^{n} Z_t \exp(-i2\pi xt) + V_{n,j} \right)$$

$$= \psi(x) J_{n,Z}(x) + Y_n(x), \tag{10.13}$$

where

$$V_{n,j} = \sum_{t=1-j}^{n-j} Z_t \exp(-i2\pi xt) - \sum_{t=1}^{n} Z_t \exp(-i2\pi xt),$$

$$Y_n(x) = \frac{1}{b_n} \sum_{j=-\infty}^{\infty} a_j \exp(-i2\pi xj) V_{n,j}.$$

Since f_X is bounded away from 0 and (10.13) holds, it is enough to show that $\max_{1 \le k \le n} |Y_n(k/n)| \overset{P}{\to} 0$. Now

$$Y_n(x) = \frac{1}{b_n} \sum_{j=n+1}^{\infty} a_j \exp(-i2\pi xj) V_{n,j} + \frac{1}{b_n} \sum_{j=1}^{n} a_j \exp(-i2\pi xj) V_{n,j}$$

$$+ \frac{1}{b_n} \sum_{j=-\infty}^{-n-1} a_j \exp(-i2\pi xj) V_{n,j} + \frac{1}{b_n} \sum_{j=-n}^{-1} a_j \exp(-i2\pi xj) V_{n,j}$$

$$= S_1(x) + S_2(x) + S_3(x) + S_4(x).$$

Now following an argument similar to that given in the proof of Lemma 2.6 in Mikosch et al. (2000), we can show that

$$\max_{1 \le k \le n} |S_i(k/n)| \overset{P}{\to} 0 \quad \text{for} \quad i = 1, 2.$$

The behavior of $S_3(x)$ and $S_4(x)$ are similar to $S_1(x)$ and $S_2(x)$, respectively. For the sake of completeness we give the details.

Note that

$$S_1(x) = -J_{n,Z}(x) \sum_{j=n+1}^{\infty} a_j \exp(-i2\pi xj)$$

$$+ \frac{1}{b_n} \sum_{j=n+1}^{\infty} a_j \exp(-i2\pi xj) \sum_{t=1-j}^{n-j} Z_t \exp(-i2\pi xj)$$

$$= S_{11}(x) + S_{12}(x),$$

and

$$\max_{1 \le k \le n} |S_{11}(k/n)| \le \max_{1 \le k \le n} |J_{n,Z}(x)| \sum_{j=n+1}^{\infty} |a_j| \overset{P}{\to} 0.$$

To show that $\max_{1\leq k\leq n} |S_{12}(k/n)| \overset{P}{\to} 0$, we write

$$|S_{12}(x)| \leq \frac{1}{b_n} \sum_{t=-n}^{-1} |Z_t| \sum_{n+1}^{n-t} |a_j| + \frac{1}{b_n} \sum_{t=-\infty}^{-n-1} |Z_t| \sum_{j=1-t}^{n-t} |a_j|. \tag{10.14}$$

Note that

$$\sum_{t=-n}^{-1} \frac{|Z_t|}{b_n} \sum_{j=n+1}^{n-t} |a_j| \overset{D}{=} \sum_{l=1}^{n} \frac{|Z_l|}{b_n} \sum_{j=n+1}^{n+l} |a_j| \overset{P}{\to} 0,$$

since $\frac{1}{b_n}\sum_{l=1}^{n}|Z_l| \overset{D}{\to} Y_\alpha$ (see Feller (1971)), and by (10.12),

$$\sum_{j=n+1}^{\infty} |a_j| \to 0.$$

To tackle the second term in (10.14), we observe that

$$\phi_n(\lambda) := \mathrm{E}\Big(\exp\Big\{-\lambda\frac{1}{b_n}\sum_{t=-\infty}^{-n-1}|Z_t|\sum_{j1-t}^{n-t}|a_j|\Big\}\Big) = \mathrm{E}[\exp(T_1 + \cdots + T_n)],$$

$$\tag{10.15}$$

where $T_j = -\lambda\frac{1}{b_n}\sum_{i=n+1}^{\infty}|Z_i||a_{i+j}|$, $j = 1,\ldots,n$. Note that

$$\mathrm{E}\exp(T_1 + \cdots + T_n) \geq \prod_{j=1}^{n} \mathrm{E}\exp(T_j) \geq \big(\mathrm{E}\exp(T_1)\big)^n, \tag{10.16}$$

since $T_1,\ldots,T_n$ are associated. Let $\phi(\lambda) = \mathrm{E}\exp\{-\lambda|Z_1|\}$. Now by Karamata's Tauberian theorem (see page 471, Feller (1971)),

$$-\log\phi(\lambda) \sim 1 - \phi(\lambda) \sim \Gamma(1-\alpha)\lambda^\alpha L(1/\lambda), \quad \lambda\downarrow 0. \tag{10.17}$$

where $L(tx)/L(t) \to 1$ as $t \to \infty$. Now using the Potter bounds (see Proposition 0.8 in Resnick (1987)), we see that for large n and some $c > 0$,

$$\begin{aligned}
\mathrm{E}\exp(T_1) &= \prod_{i=n+1}^{\infty} \phi(\frac{1}{b_n}\lambda|a_j|) \\
&= \exp\Big\{-\sum_{i=n+2}^{\infty}(-\log\phi(\frac{1}{b_n}\lambda|a_j|))\Big\} \\
&\geq \exp\Big\{-\frac{c}{n}\sum_{i=n+2}^{\infty}\frac{-\log\phi(\frac{1}{b_n}\lambda|a_j|)}{-\log\phi(\frac{1}{b_n}\lambda)}\Big\} \\
&\geq \exp\Big\{-\frac{c}{n}\sum_{i=n+2}^{\infty}|a_j|^{(\alpha-\varepsilon)}\Big\}, \quad \text{by } (10.17),
\end{aligned}$$

and hence from (10.15) and (10.16), we conclude that $\phi_n(\lambda) \to 1$. Hence

$$\max_{1 \le k \le n} |S_{12}(k/n)| \xrightarrow{P} 0.$$

Now we write S_2 in the following way:

$$
\begin{aligned}
|S_2(x)| &= \Big| \frac{1}{b_n} \sum_{j=1}^{n} a_j \exp(-i2\pi x j) \sum_{t=1-j}^{0} Z_t \exp(-i2\pi x t) \\
&\quad - \frac{1}{b_n} \sum_{j=1}^{n} a_j \exp(-i2\pi x j) \sum_{t=n-j+1}^{n} Z_t \exp(-i2\pi x t) \Big| \\
&\le \frac{1}{b_n} \sum_{j=1}^{n} |a_j| \Big(\sum_{t=1-j}^{0} |Z_t| + \sum_{t=n-j+1}^{n} |Z_t| \Big) \\
&= S_{21} + S_{22}.
\end{aligned}
$$

Now

$$
\begin{aligned}
S_{21} &= \sum_{j=1}^{n} |a_j| \sum_{t=1-j}^{0} \frac{|Z_t|}{b_n} \\
&\stackrel{\mathcal{D}}{=} \sum_{j=1}^{n} |a_j| \sum_{k=1}^{j} \frac{|Z_k|}{b_n} \\
&= \sum_{k=1}^{k_0} \Big(\sum_{j=k}^{n} |a_j| \Big) \frac{|Z_k|}{b_n} + \sum_{k=k_0+1}^{n} \Big(\sum_{j=k}^{n} |a_j| \Big) \frac{|Z_k|}{b_n} \\
&= S_{211} + S_{212},
\end{aligned}
$$

and for fixed k_0,

$$S_{211} \le \Big(\sum_{k=1}^{k_0} \frac{|Z_k|}{b_n} \Big) \sum_{j=1}^{\infty} |a_j| \xrightarrow{P} 0.$$

If we choose k_0 such that $\sum_{k=k_0}^{\infty} |a_k| < \varepsilon$, then

$$S_{212} \le \varepsilon \sum_{k=k_0+1}^{n} \frac{|Z_k|}{b_n} \le \varepsilon \sum_{k=1}^{n} \frac{|Z_k|}{b_n} \xrightarrow{\mathcal{D}} \varepsilon Y_\alpha.$$

Since $\varepsilon > 0$ can be chosen arbitrarily small, we conclude that $S_{212} \xrightarrow{P} 0$. This proves that $S_{21} \xrightarrow{P} 0$. For S_{22}, we note that

$$b_n S_{22} = \sum_{j=1}^{n} |a_j| \sum_{t=n-j+1}^{n} |Z_t| = \sum_{t=1}^{n} |Z_t| \sum_{j=n-t+1}^{n} |a_j| \stackrel{\mathcal{D}}{=} \sum_{t=1}^{n} |Z_t| \sum_{j=t}^{n} |a_j|.$$

Now the expression on the right side of the last equation can be analyzed in a similar way as S_{21}. This completes the proof of the fact that

$$\max_{1 \leq k \leq n} |S_2(k/n)| \overset{\mathcal{P}}{\to} 0.$$

The behavior of $S_3(x)$ and $S_4(x)$ are similar to $S_1(x)$ and $S_2(x)$ respectively. Therefore, following a similar argument we can show that $\max_{1 \leq k \leq n} |S_j(k/n)| \overset{\mathcal{P}}{\to} 0$ for $j = 3, 4$. This completes the proof of part (a).

(b) Let $\widehat{SC}_n$ be the symmetric circulant matrix formed with independent entries $\{Z_i\}$. In view of Theorem 10.3.1, it is enough to show that

$$\left| M\left(\frac{1}{b_n} SC_n, f_X\right) - \mathrm{sp}\left(\frac{1}{b_n} \widehat{SC}_n\right) \right| \overset{\mathcal{P}}{\to} 0.$$

Rest of the proof is similar to part (a). We leave it as an exercise. □

10.5   Exercises

1. Show that

$$\sum_{j=1}^{\infty} \Gamma_j^{-1/\alpha} \overset{\mathcal{D}}{\simeq} S_\alpha(C_\alpha^{-\frac{1}{\alpha}}, 1, 0) \quad \text{where} \quad C_\alpha = \left(\int_0^{\infty} x^{-\alpha} \sin x \, dx \right)^{-1}.$$

2. Prove Theorem 10.3.1 when n is even.

3. (Exercise 0.4.2.8 of Resnick (1987)) Suppose $F(x)$ is a distribution on $\mathbb{R}_+$ and

$$1 - F(x) \sim x^{-\alpha} L(x),$$

where L is a slowly varying function. Using integration by parts or Fubini's theorem, show that for $\eta \geq \alpha$,

$$\lim_{x \to \infty} \frac{\int_0^x u^\eta F(du)}{x^\eta (1 - F(x))} = \frac{\alpha}{\eta - \alpha}.$$

4. Complete the proof of part (b) of Theorem 10.4.1.

11

Appendix

The purpose of this Appendix is three-fold: first, to give a detailed proof of the eigenvalue formula solution for the k-circulant that has been used crucially in the book; second, to briefly define and explain some of the background material in probability theory that was needed in the main text; finally, to state a few technical results from the literature that have been used repeatedly in the proofs of our results.

11.1   Proof of Theorem 1.4.1

In what follows, $\chi(A)(\lambda)$ stands for the characteristic polynomial of the matrix A evaluated at λ but for ease of notation, we shall often suppress the argument λ and write simply $\chi(A)$.

We recall $\{\alpha_q\}$ and $\{\beta_q\}$ from (1.4) and define

$$m := \max_{1 \le q \le c} \lceil \beta_q / \alpha_q \rceil, \quad [t]_{m,b} := tk^m \bmod b, \ b \text{ is a positive integer}. \quad (11.1)$$

Let $e_{m,d}$ be a $d \times 1$ vector whose only non-zero element is 1 at $(m \bmod d)$-th position, $E_{m,d}$ be the $d \times d$ matrix with $e_{jm,d}$, $0 \le j < d$ as its columns, and for dummy symbols $\delta_0, \delta_1, \ldots$, let $\Delta_{m,b,d}$ be a diagonal matrix as given below.

$$
\begin{aligned}
e'_{m,d} &= [0 \ \cdots \ 0 \ 1 \ 0 \ \cdots \ 0]_{1 \times d}, & (11.2) \\
E_{m,d} &= [e_{0,d} \ e_{m,d} \ e_{2m,d} \cdots e_{(d-1)m,d}], & (11.3) \\
\Delta_{m,b,d} &= \operatorname{diag} [\delta_{[0]_{m,b}}, \ \delta_{[1]_{m,b}}, \ \ldots, \ \delta_{[j]_{m,b}}, \ \ldots, \ \delta_{[d-1]_{m,b}}]. & (11.4)
\end{aligned}
$$

Note that

$$\Delta_{0,b,d} = \operatorname{diag} [\delta_{0 \bmod b}, \ \delta_{1 \bmod b}, \ \ldots, \ \delta_{j \bmod b}, \ \ldots, \ \delta_{d-1 \bmod b}].$$

We need the following lemmata for the main proof.

Lemma 11.1.1. Let $\pi = (\pi(0), \ \pi(1), \ \ldots, \ \pi(b-1))$ be a permutation of $(0, 1, \ldots, b-1)$. Let

$$P_\pi = [e_{\pi(0),b} \ e_{\pi(1),b} \cdots e_{\pi(b-1),b}].$$

Then, P_π is a permutation matrix and the (i, j)th element of $P_\pi^T E_{k,b} \Delta_{0,b,b} P_\pi$ is given by

$$(P_\pi^T E_{k,b} \Delta_{0,b,b} P_\pi)_{i,j} = \begin{cases} \delta_t & \text{if } (i, j) = (\pi^{-1}(kt \bmod b), \pi^{-1}(t)), \quad 0 \le t < b \\ 0 & \text{otherwise.} \end{cases}$$

The proof is easy and we omit it.

Lemma 11.1.2. Let k and b be positive integers. Then

$$\chi(A_{k,b}) = \chi(E_{k,b} \Delta_{0,b,b}),$$

where $\delta_j = \sum_{l=0}^{b-1} a_l \omega^{jl}$, $0 \le j < b$, $\omega = \cos(2\pi/b) + i \sin(2\pi/b)$, $i = \sqrt{-1}$.

Proof. Define the $b \times b$ permutation matrix

$$P_b = \begin{bmatrix} \underline{0} & I_{b-1} \\ 1 & \underline{0}^T \end{bmatrix}.$$

Observe that for $0 \le j < b$, the j-th row of $A_{k,b}$ can be written as $a^T P_b^{jk}$ where P_b^{jk} stands for the jk-th power of P_b. From direct calculation, it is easy to verify that $P_b = UDU^*$ is a spectral decomposition of P_b, where

$$\begin{aligned} D &= \text{diag}(1, \omega, \ldots, \omega^{b-1}), & (11.5) \\ U &= [u_0 \ u_1 \ \cdots \ u_{b-1}] \text{ with} & (11.6) \\ u_j &= b^{-1/2}(1, \omega^j, \omega^{2j}, \ldots, \omega^{(b-1)j})^T, \quad 0 \le j < b. \end{aligned}$$

Note that $\delta_j = a^T u_j$, $0 \le j < b$. From easy computations, it now follows that

$$U^* A_{k,b} U = E_{k,b} \Delta_{0,b,b},$$

so that, $\chi(A_{k,b}) = \chi(E_{k,b} \Delta_{0,b,b})$, and the lemma is proven. $\qquad\square$

Lemma 11.1.3. Let k and b be positive integers, and $x = b/\gcd(k, b)$. Let for dummy variables $\gamma_0, \gamma_1, \gamma_2, \ldots, \gamma_{b-1}$,

$$\Gamma = \text{diag}(\gamma_0, \gamma_1, \gamma_2, \ldots, \gamma_{b-1}).$$

Then

$$\chi(E_{k,b} \times \Gamma) = \lambda^{b-x} \chi\left(E_{k,x} \times \text{diag}\left(\gamma_{0 \bmod b}, \gamma_{k \bmod b}, \ldots, \gamma_{(x-1)k \bmod b}\right)\right).$$
$$(11.7)$$

Proof. Define the matrices,

$$B_{b \times x} = \begin{bmatrix} e_{0,b} & e_{k,b} & e_{2k,b} & \cdots & e_{(x-1)k,b} \end{bmatrix} \text{ and } P = [B \ B^c],$$

where B^c consists of those columns (in any order) of I_b that are not in B. This makes P a permutation matrix.

Clearly, $E_{k,b} = [B \; B \; \cdots \; B]$ which is a $b \times b$ matrix of rank x, and we have

$$\chi\left(E_{k,b}\Gamma\right) = \chi\left(P^T E_{k,b}\Gamma P\right).$$

Note that

$$P^T E_{k,b}\Gamma P \;=\; \begin{bmatrix} I_x & I_x & \cdots & I_x \\ 0_{(b-x)\times x} & 0_{(b-x)\times x} & \cdots & 0_{(b-x)\times x} \end{bmatrix} \Gamma P$$

$$=\; \begin{bmatrix} C \\ 0_{(b-x)\times b} \end{bmatrix} P$$

$$=\; \begin{bmatrix} C \\ 0_{(b-x)\times b} \end{bmatrix} [B \; B^c] = \begin{bmatrix} CB & CB^c \\ 0 & 0 \end{bmatrix},$$

where

$$C \;=\; [I_x \; I_x \; \cdots \; I_x]\,\Gamma$$

$$=\; [I_x \; I_x \; \cdots \; I_x] \times \mathrm{diag}(\gamma_0, \; \gamma_1, \; \ldots, \; \gamma_{b-1}).$$

Clearly, the characteristic polynomial of $P^T E_{k,b}\Gamma P$ does not depend on CB^c. This explains why we had not bothered to specify the order of columns in B^c. Thus we have

$$\chi\left(E_{k,b}\Gamma\right) = \chi\left(P^T E_{k,b}\Gamma P\right) = \lambda^{b-x}\chi\left(CB\right).$$

It now remains to show that

$$CB = E_{k,x} \times \mathrm{diag}\left(\gamma_{0 \bmod b}, \; \gamma_{k \bmod b}, \; \gamma_{2k \bmod b}, \cdots, \gamma_{(x-1)k \bmod b}\right).$$

Note that the j-th column of B is $e_{jk,b}$. So the j-th column of CB is actually the $(jk \bmod b)$-th column of C. Hence $(jk \bmod b)$-th column of C is $\gamma_{jk \bmod b}\, e_{jk \bmod x}$. So

$$CB = E_{k,x} \times \mathrm{diag}\left(\gamma_{0 \bmod b}, \; \gamma_{k \bmod b}, \; \gamma_{2k \bmod b}, \cdots, \gamma_{(x-1)k \bmod b}\right),$$

and the Lemma is proven completely.   □

Proof of Theorem 1.4.1.. We first prove the Theorem for $A_{k,n'}$. Since k and n' are relatively prime, by Lemma 11.1.2,

$$\chi(A_{k,n'}) = \chi(E_{k,n'}\Delta_{0,n',n'}).$$

Get the partitioning sets $\mathcal{P}_0$, $\mathcal{P}_1$, $\ldots$ of $\{0,1,\ldots,n'-1\}$, as in (1.7) where $\mathcal{P}_j = \{r_j k^x \bmod n', \; 0 \le x < \#\mathcal{P}_j\}$ for some integer r_j. Let $N_0 = 0$ and $N_j = \sum_{i=1}^{j} n_i$ where $n_i = \#\mathcal{P}_i$. Define a permutation π on the set $\mathbb{Z}_{n'}$ as follows:

$$\pi(0) = 0 \text{ and } \pi(N_j + t) = r_{j+1}k^{t-1} \quad \bmod n' \text{ for } 1 \le t \le n_{j+1} \text{ and } j \ge 0.$$

This permutation π automatically yields a permutation matrix P_π as in Lemma 11.1.1. Consider the positions of δ_v for $v \in \mathcal{P}_j$ in the product $P_\pi^T E_{k,n'} \Delta_{0,n',n'} P_\pi$. We know that $v = r_j k^{t-1} \mod n'$ for some $1 \le t \le n_j$. Thus

$$\pi^{-1}\left(r_j k^{t-1} \mod n'\right) = N_{j-1} + t, \quad 1 \le t \le n_j,$$

so that the position of δ_v for $v = r_j k^{t-1} \mod n'$, $1 \le t \le n_j$ in $P_\pi^T E_{k,n'} \Delta_{0,n'} P_\pi$ is given by,

$$\left(\pi^{-1}(r_j k^t \mod n'), \pi^{-1}(r_j k^{t-1} \mod n')\right)$$
$$= \begin{cases} (N_{j-1} + t + 1, \ N_{j-1} + t) & \text{if, } 1 \le t < n_j \\ (N_{j-1} + 1, \ N_{j-1} + n_j) & \text{if, } t = n_j \end{cases}.$$

Hence

$$P_\pi^T E_{k,n'} \Delta_{0,n',n'} P_\pi = \text{diag}\left(L_0, \ L_1, \ \dots\right),$$

where for $j \ge 0$, if $n_j = 1$ then $L_j = \left[\delta_{r_j}\right]$ is a 1×1 matrix, and if $n_j > 1$, then

$$L_j = \begin{bmatrix} 0 & 0 & 0 & \cdots & 0 & \delta_{r_j k^{n_j - 1} \mod n'} \\ \delta_{r_j \mod n'} & 0 & 0 & \cdots & 0 & 0 \\ 0 & \delta_{r_j k \mod n'} & 0 & \cdots & 0 & 0 \\ & & & \vdots & & \\ 0 & 0 & 0 & \cdots & \delta_{r_j k^{n_j - 2} \mod n'} & 0 \end{bmatrix}.$$

Clearly, $\chi(L_j) = \lambda^{n_j} - y_j$. Now the result follows from the identity

$$\chi\left(E_{k,n'}\Delta_{0,n',n'}\right) = \prod_{j \ge 0} \chi(L_j) = \prod_{j \ge 0}(\lambda^{n_j} - y_j).$$

Now let us prove the result for the general case. Recall that $n = n' \times \prod_{q=1}^c p_q^{\beta_q}$. Then again using Lemma 11.1.2,

$$\chi(A_{k,n}) = \chi(E_{k,n}\Delta_{0,n,n}).$$

Recalling (11.1) and Lemma 11.1.2 and using Lemma 11.1.3 repeatedly with $y = n/n'$,

$$\begin{aligned} \chi(A_{k,n}) &= \chi(E_{k,n}\Delta_{0,n,n}) \\ &= \lambda^{n-n'}\chi(E_{k,n'}\Delta_{m,n,n'}) \\ &= \lambda^{n-n'}\chi(E_{k,n'}\Delta_{m+j,n,n'}) \quad [\text{for all } j \ge 0] \\ &= \lambda^{n-n'}\chi\left(E_{k,n'} \times \text{diag}\left(\delta_{[0]_{0,n}}, \ \delta_{[y]_{0,n}}, \ \delta_{[2y]_{0,n}}, \dots, \delta_{[(n'-1)y]_{0,n}}\right)\right). \end{aligned}$$

Replacing $\Delta_{0,n',n'}$ by $\text{diag}\left(\delta_{[0]_{0,n}}, \ \delta_{[y]_{0,n}}, \ \delta_{[2y]_{0,n}}, \dots, \delta_{[(n'-1)y]_{0,n}}\right)$, we can mimic the rest of the proof given for $A_{k,n'}$, to complete the proof in the general case. $\qquad\square$

11.2   Standard notions and results

Probability inequalities: In this book we have used a few elementary probability inequalities.

Boole's inequality: Suppose $\{E_i, i \geq 1\}$ are events in some probability space $(\Omega, \mathcal{A}, P)$. Then

$$P\left(\bigcup_{n=1}^{\infty} E_i\right) \leq \sum_{n=1}^{\infty} P(E_i). \tag{11.8}$$

Bonferroni inequality: These provide both upper and lower bounds, and are based on the inclusion-exclusion principle. Suppose $E_1, E_2, \ldots, E_n$ are events in some probability space $(\Omega, \mathcal{A}, P)$. Then for every integer m such that $2 \leq 2m \leq n$,

$$\sum_{j=1}^{2m}(-1)^{j-1}S_{j,n} \leq P\left(\bigcup_{j=1}^{n} E_i\right) \leq \sum_{j=1}^{2m-1}(-1)^{j-1}S_{j,n},$$

where

$$S_{j,n} := \sum_{1 \leq i_1 < i_2 < \cdots < i_j \leq n} P\left(\bigcap_{l=1}^{j} E_{i_l}\right).$$

Chebyshev's inequality: This bounds the probability for deviation about the mean when the variance is known to be finite. Suppose X is a real-valued random variable with mean μ and variance σ^2. Then

$$P(|X - \mu| \geq x) \leq \frac{\sigma^2}{x^2}, \quad \text{for any } x > 0. \tag{11.9}$$

Holder's inequality: Suppose X and Y are real-valued random variables. Suppose p, q are positive numbers. Then

$$E\,|XY| \leq [E\,|X^p|]^{\frac{1}{p}}[E\,|X^q|]^{\frac{1}{q}}, \quad \text{whenever } \frac{1}{p} + \frac{1}{q} = 1.$$

The special case $p = q = 2$ is known as Cauchy-Schwarz inequality.

Bernstein's inequality: Let $X_1, X_2, \ldots, X_n$ be independent real-valued bounded random variables such that $E\,X_i = 0$ and $|X_i| < M$ almost surely, for $1 \leq i \leq n$. Then for any $t > 0$,

$$P\left(\frac{1}{n}\sum_{i=1}^{n} X_i \geq t\right) \leq \exp\left(-\frac{\frac{1}{2}t^2}{\sum_{i=1}^{n} E(X_i^2) + \frac{1}{3}Mt}\right).$$

Convergence notions: The three fundamental notions of convergence in probability theory are convergence in distribution, convergence in probability and almost sure convergence.

Convergence in distribution: Suppose $\{X_n\}$ is a sequence of $\mathbb{R}^d$-valued random variables with distribution functions $\{F_n\}$ and X is an $\mathbb{R}^d$-valued random variable with distribution function F. Then X_n is said to converge in distribution to X if

$$F_n(x) \to F(x) \text{ for every continuity point } x \in \mathbb{R}^d \text{ of } F.$$

Note that X_n's, and X too, may be defined on different probability spaces.

Convergence in probability: Suppose $\{X_n\}$ is a sequence of $\mathbb{R}^d$-valued random variables and X is another $\mathbb{R}^d$-valued random variable, all defined on the same probability space. Then we say that X_n converges to X *in probability* if

$$P(\|X_n - X\| \geq x) \to 0 \text{ for every } x > 0.$$

Often the convergence is verified by using Chebyshev's inequality. Convergence in probability implies convergence in distribution.

Almost sure convergence: Suppose $\{X_n\}$ is a sequence of $\mathbb{R}^d$-valued random variables and X is another $\mathbb{R}^d$-valued random variable, all defined on the same probability space. Then we say that X_n converges to X *almost surely* if

$$P\{\omega : \lim_{n \to \infty} X_n(\omega) = X(\omega)\} = 1.$$

Almost sure convergence implies convergence in probability.

Borel-Cantelli lemma: Suppose $\{E_n\}$ is a sequence of events in a probability space. Then

$$\sum_{n=1}^{\infty} P(E_n) < \infty \text{ implies that } P(\limsup E_n) = P(\cap_{n=1}^{\infty} \cup_{k=n}^{\infty} E_k) = 0.$$

Dominated Convergence Theorem (DCT): Suppose $\{X_n\}$ is a sequence of real-valued random variables which converges almost surely to X. If there exists a random variable Y such that $\sup_n |X_n| \leq Y$ and $EY < \infty$, then $E(X_n) \to E(X)$.

Limit theorems for sums: Three basic limit theorems in probability are the strong law of large numbers (SLLN), the central limit theorem (CLT) and the law of iterated logarithm.

Central Limit Theorem: This is the first fundamental limit theorem in probability theory and is an excellent non-trivial example of convergence in distribution. Suppose $\{X_n\}$ is a sequence of i.i.d. real-valued random variables defined on the same probability space with mean μ and finite variance σ^2. Then

$$\frac{X_1 + X_2 + \cdots + X_n - n\mu}{\sqrt{n}\sigma} \to N(0,1) \text{ in distribution,}$$

where $N(0, 1)$ denotes the standard normal distribution.

Strong Law of Large Numbers: This is the second fundamental limit theorem in probability theory and is an excellent non-trivial example of almost sure convergence. Suppose $\{X_n\}$ is a sequence of i.i.d. real-valued random variables on the same probability space with mean μ. Then

$$\frac{X_1 + X_2 + \cdots + X_n}{n} \to \mu \quad \text{almost surely.}$$

Law of Iterated Logarithm: This is the third fundamental limit theorem in probability theory. Suppose $\{X_n\}$ is a sequence of i.i.d. real-valued random variables with $\mathrm{E}\, X_i = 0$ and $\mathrm{Var}(X_i) = 1$. Then

$$\limsup \frac{\sum_{i=1}^n X_i}{\sqrt{2n \ln \ln n}} = - \liminf \frac{\sum_{i=1}^n X_i}{\sqrt{2n \ln \ln n}} = 1 \quad \text{almost surely.}$$

Normal approximation: As in the Central Limit Theorem, there are many situations where the limit distribution is normal. When this is the case, one tries to provide the accuracy of the normal approximation. The simplest but a far reaching result in this direction is the following bound.

Berry-Esseen Bound: Suppose $\{X_n\}$ is a sequence of i.i.d. real-valued random variables defined on the same probability space with mean μ and finite variance σ^2. Then

$$\sup_{x \in \mathbb{R}} \left| \mathrm{P}\Big(\frac{X_1 + X_2 + \cdots + X_n - n\mu}{\sqrt{n}\sigma} \le x \Big) - \mathrm{P}(X \le x) \right| \le \frac{C\, \mathrm{E}\, |X_1|^3}{\sigma^3 \sqrt{n}}$$

where C is a universal constant independent of the distribution of X_1 and X is a normal random variable with mean zero and variance 1. In the main text, we have used other sophisticated normal approximation results from the literature.

Stationary processes: Broadly speaking, stationary processes are those whose probabilistic behavior (in some sense) remains invariant over time. The basic notions we have used are the following:

Weakly stationary sequence: A sequence of real-valued random variables $\{X_n\}$ is said to be *weakly stationary* if $\mathrm{E}(X_n)$ does not depend on n and there exists a function $\gamma(\cdot)$ on non-negative integers such that $Cov(X_n, X_m) = \gamma(|n - m|)$ for all values of n and m.

Spectral distribution and density: Any such covariance sequence $\gamma(\cdot)$ is associated with a (unique) distribution function (non-decreasing, right-continuous) F with support on $[0, 2\pi]$ and with the property that

$$\gamma(k) = \int_0^{2\pi} e^{ikx} dF(x) \quad \text{for all } k.$$

If F has a density f, then it is called the *spectral density* of the process $\{X_n\}$. It is well-known that if $\sum |\gamma(n)| < \infty$, then the spectral density exists.

White noise: If $\gamma(n) = 0$ for all $n \neq 0$, Then $\{X_n\}$ is called a *white noise*. It is clear that the spectral density of a white noise sequence is given by

$$f(x) = \frac{\gamma(0)}{2\pi}, \text{ for all } x \in [0, 2\pi].$$

Extreme value theory: There is a well-developed theory of the probabilistic behavior of ordered and extreme values of random sequences. We have borrowed the following notions from this theory:

Convergence of types theorem: This is a technical result which allows the replacement of normalizing constants by some equivalent sequence. Let X_n be a sequence of real-valued random variables which converges in distribution to X. Suppose further that for some sequences $a_n > 0$ and $b_n \in \mathbb{R}$, $a_n X_n + b_n$ converges in distribution to X' and both X and X' are non-degenerate. Then $a_n \to a$ and $b_n \to b$ for some $a > 0$, $b \in \mathbb{R}$. Equivalently, suppose $\{F_n\}$, F and F' are distribution functions where F and F' are non-degenerate. Moreover, there exist $a_n, a'_n > 0$ and $b_n, b'_n \in \mathbb{R}$ such that $F_n(a_n x + b_n) \to F(x)$ and $F_n(a'_n x + b'_n) \to F'(x)$ for all continuity points x of F, respectively F'. Then $a_n/a'_n \to a > 0$, and $(b_n - b'_n)/a'_n \to b \in \mathbb{R}$, and $F'(ax + b) = F(x)$ for all $x \in \mathbb{R}$.

Standard Gumbel distribution: Its distribution function is defined as

$$\Lambda(x) = \exp\{-\exp(-x)\}, \ x \in \mathbb{R}.$$

The Gumbel distribution arises in the study of limit distribution of the maximum. Suppose $\{X_n\}$ is a sequence of i.i.d. real-valued random variables with distribution function F. Let $M_n = \max\{X_1, \ldots, X_n\}$. In a fundamental result in extreme value theory, necessary sufficient conditions are given for the distributional convergence of $M_n^* = (M_n - b_n)/a_n$ to a non-degenerate limit for suitable choices of $\{a_n, b_n\}$, and three classes of limit distributions are possible. The one which is relevant to us is the Gumbel distribution. In that case we say that X_1 or its distribution F is in the *max-domain of attraction* of the Gumbel distribution. The numbers a_n and b_n are called the norming constants. Here is the result that we have used.

A distribution function $F_\#$ with right end x_0 (can be ∞) is called a *Von Mises function* if it has the following representation: there must exist $z_0 < x_0$ such that for $z_0 < x < x_0$ and $c > 0$,

$$1 - F_\#(x) = c \exp\left\{-\int_{z_0}^{x} (1/f(u)) du\right\}, \tag{11.10}$$

where $f(u) > 0$ for $z_0 < x < x_0$, and f is absolutely continuous on (z_0, x_0) with density $f'(u)$ and $\lim_{u\uparrow x_0} f'(u) = 0$.

Proposition 11.2.1. If $F_\#$ is a Von Mises function with representation

(11.10) then $F_{\#}$ belongs to the max-domain of attraction of the Gumbel distribution. The norming constants may be chosen as

$$b_n = (1/(1-F))^{\leftarrow}(n) \quad \text{and} \quad a_n = f(b_n). \tag{11.11}$$

Here for any non-decreasing function G,

$$G^{\leftarrow}(x) = \inf\{y : G(y) \ge x\}.$$

There are criteria which help to identify the norming constants a_n and b_n. The convergence of types result also comes in handy in picking these constants. For instance, if F_1 is the standard normal distribution, it is known that in (11.11), a_n and b_n can be taken to be

$$b_n = (2\ln n - \ln\ln n - \ln(4\pi))^{1/2} \quad \text{and} \quad a_n = \frac{1}{b_n}.$$

Heavy-tailed distributions: Variables with infinite second moment, and in general with heavy tails, have very different properties compared to the light-tailed ones. We have used the following notions:

Stable distributions: Suppose X is a random variable. Let X_1 and X_2 be independent random variables with the same distribution as that of X. Then X or its distribution is said to be *stable* if for any constants $a, b > 0$ the random variable $aX_1 + bX_2$ has the same distribution as $cX + d$ for some constants $c > 0$ and d. For example the normal and the Cauchy distributions are stable.

Regularly varying function: A measurable function $U : \mathbb{R}_+ \to \mathbb{R}_+$ is said to be regularly varying at ∞ with index ρ if for $x > 0$,

$$\lim_{t\to\infty} \frac{U(tx)}{U(t)} = x^{\rho}.$$

We denote this class of function as RV_{ρ}.

If $\rho = 0$ we call U *slowly varying*. Slowly varying functions are generally denoted by $L(x)$ or $\ell(x)$. Note that if $U \in RV_{\rho}$ then $\frac{U(x)}{x^{\rho}} \in RV_0$.

Notions from Stochastic Processes

Vague convergence: A sequence of measures μ_n is said to converge vaguely to another measure μ, if for every bounded continuous function f with compact support, $\int f(x)d\mu_n(x) \to \int f(x)d\mu(x)$.

Point process: Roughly speaking, a *point process* is a collection of points randomly located on some suitable space. Formally, let E be any set equipped with an appropriate sigma-algebra $\mathcal{E}$. For example usually E is some nice subset of $\mathbb{R}^d$ and is equipped with the Borel sigma-algebra. Let $M(E)$ be the space of all point measures on E, endowed with the topology of vague convergence. Any *point process* on E is a measurable map

$$N : (\Omega, \mathcal{F}, \mathrm{P}) \to (M(E), \mathcal{E}).$$

It is said to be *simple* if

$$P(N(\{x\}) \leq 1, x \in E) = 1.$$

Poisson point process: The simplest example of a point process is a *Poisson point process*. Let $\lambda(x)$, $x \in \mathbb{R}^d$ be a locally integrable positive function so that

$$\Lambda(B) := \int_B \lambda(x)dx < \infty \quad \text{for any bounded region } B.$$

Then $N(\cdot)$ is said to be a Poisson point process with intensity function $\lambda(\cdot)$ if for any collection of disjoint bounded Borel measurable sets $B_1, B_2, \ldots, B_k$:

$$P(N(B_i) = n_i, \ i = 1, 2, \ldots, k) = \prod_{i=1}^{k} \frac{(\Lambda(B_i))^{n_i}}{n_i!} e^{-\Lambda(B_i)}.$$

Clearly, any Poisson point process is simple.

11.3   Three auxiliary results

The following three results have been repeatedly used in the text.

Theorem 11.3.1 (Theorem 7.1.2; Brockwell and Davis (2006)). Suppose $\{X_t\}$ satisfies

$$X_t = \mu + \sum_{j=-\infty}^{\infty} \psi_j Z_{t-j}, \quad \{Z_t\} \sim \text{IID}(0, \sigma^2),$$

where $\sum_{j=-\infty}^{\infty} |\psi_j| < \infty$ and $\sum_{j=-\infty}^{\infty} \psi_j \neq 0$. Then

$$n^{1/2} v^{-1/2} (\bar{X}_n - \mu) \xrightarrow{D} N(0, 1),$$

where $\bar{X}_n = \frac{1}{n} \sum_{i=1}^{n} X_i$, $v = \sum_{h=-\infty}^{\infty} \gamma(h) = \sigma^2 (\sum_{j=-\infty}^{\infty} \psi_j)^2$, and $\gamma(\cdot)$ is the autocovariance function of $\{X_t\}$.

Theorem 11.3.2 (Theorem 2.1; Davis and Mikosch (1999)). Let $\{Z_t\}$ be a sequence of real-valued i.i.d. random variables with mean zero and variance one. If $E|Z_1|^s < \infty$ for some $s > 2$, then

$$\max_{k=1,\ldots,q_n} I_{n,Z}(\omega_k) - \ln q_n \xrightarrow{D} Y,$$

where

$$I_{n,Z}(\omega_i) = n^{-1} \left| \sum_{j=1}^{n} \exp(-i\omega_k j) Z_j \right|^2, \quad \omega_k = \frac{2\pi k}{n},$$

$$q_n = \max\{k : 0 < \omega_k < \pi\}, \text{ that is, } q_n \sim \frac{n}{2}, \text{ and}$$

Y has the standard Gumbel distribution $\Lambda(x) = \exp(-\exp(-x))$, $x \in \mathbb{R}$.

Karamata's theorem examines the properties of the integral of regularly varying functions. Suppose U is locally integrable and is also integrable in a neighborhood of 0. We are interested only in the behavior of functions near ∞.

Theorem 11.3.3 (Karamata's Theorem 0.6, Resnick (1987)). (a) Suppose $\rho \geq -1$ and $U \in RV_\rho$. Then

$$\int_0^x U(t)dt \in RV_{\rho+1} \quad \text{and} \quad \lim_{x\to\infty} \frac{xU(x)}{\int_0^x U(t)dt} = \rho + 1.$$

If $\rho < -1$ (or if $\rho = -1$ and $\int_x^\infty U(s)ds < \infty$), then $U \in RV_\rho$ implies that $\int_x^\infty U(t)dt$ is finite, and

$$\int_x^\infty U(t)dt \in RV_{\rho+1} \quad \text{and} \quad \lim_{x\to\infty} \frac{xU(x)}{\int_x^\infty U(t)dt} = -\rho - 1.$$

(b) If U satisfies

$$\lim_{x\to\infty} \frac{xU(x)}{\int_0^x U(t)dt} = \lambda \in (0, \infty),$$

then $U \in RV_{\lambda-1}$. If $\int_x^\infty U(t)dt < \infty$ and

$$\lim_{x\to\infty} \frac{xU(x)}{\int_x^\infty U(t)dt} = \lambda \in (0, \infty),$$

then $U \in RV_{-\lambda-1}$.

Bibliography

Auffinger, A., Ben Arous, G., and Péché, S. (2009). Poisson convergence for the largest eigenvalues of heavy tailed random matrices. *Ann. Inst. Henri Poincaré Probab. Stat.*, 45(3):589–610.

Bai, Z. D. and Silverstein, J. W. (2010). *Spectral Analysis of Large Dimensional Random Matrices*. Springer Series in Statistics. Springer, New York, second edition.

Bhatia, R. (1997). *Matrix Analysis*, volume 169 of *Graduate Texts in Mathematics*. Springer-Verlag, New York.

Bhattacharya, R. N. and Ranga Rao, R. (1976). *Normal Approximation and Asymptotic Expansions*. John Wiley & Sons, New York-London-Sydney. Wiley Series in Probability and Mathematical Statistics.

Billingsley, P. (1995). *Probability and Measure*. John Wiley & Sons, Inc., New York, third edition.

Bose, A. (2018). *Patterned Random Matrices*. Chapman & Hall.

Bose, A., Chatterjee, S., and Gangopadhyay, S. (2003). Limiting spectral distributions of large dimensional random matrices. *J. Indian Statist. Assoc.*, 41(2):221–259.

Bose, A., Guha, S., Hazra, R. S., and Saha, K. (2011a). Circulant type matrices with heavy tailed entries. *Statist. Probab. Lett.*, 81(11):1706–1716.

Bose, A., Hazra, R. S., and Saha, K. (2009). Limiting spectral distribution of circulant type matrices with dependent inputs. *Electron. J. Probab.*, 14:no. 86, 2463–2491.

Bose, A., Hazra, R. S., and Saha, K. (2010). Spectral norm of circulant type matrices with heavy tailed entries. *Electron. Commun. Probab.*, 15:299–313.

Bose, A., Hazra, R. S., and Saha, K. (2011b). Poisson convergence of eigenvalues of circulant type matrices. *Extremes*, 14(4):365–392.

Bose, A., Hazra, R. S., and Saha, K. (2011c). Spectral norm of circulant-type matrices. *J. Theoret. Probab.*, 24(2):479–516.

Bose, A., Hazra, R. S., and Saha, K. (2012a). Product of exponentials and spectral radius of random k-circulants. *Ann. Inst. Henri Poincaré Probab. Stat.*, 48(2):424–443.

Bose, A. and Mitra, J. (2002). Limiting spectral distribution of a special circulant. *Statist. Probab. Lett.*, 60(1):111–120.

Bose, A., Mitra, J., and Sen, A. (2012b). Limiting spectral distribution of random k-circulants. *J. Theoret. Probab.*, 25(3):771–797.

Bose, A. and Sen, A. (2007). Spectral norm of random large dimensional noncentral Toeplitz and Hankel matrices. *Electron. Comm. Probab.*, 12:29–35. Paging changed to 21-27 on journal site.

Bose, A. and Sen, A. (2008). Another look at the moment method for large dimensional random matrices. *Electron. J. Probab.*, 13(21):588–628.

Brockwell, P. J. and Davis, R. A. (2006). *Time Series: Theory and Methods.* Springer Series in Statistics. Springer, New York. Reprint of the second (1991) edition.

Dai, M. and Mukherjea, A. (2001). Identification of the parameters of a multivariate normal vector by the distribution of the maximum. *J. Theoret. Probab.*, 14(1):267–298.

Davis, P. J. (1979). *Circulant Matrices.* John Wiley & Sons, New York-Chichester-Brisbane.

Davis, R. A. and Mikosch, T. (1999). The maximum of the periodogram of a non-Gaussian sequence. *Ann. Probab.*, 27(1):522–536.

Einmahl, U. (1989). Extensions of results of Komlós, Major, and Tusnády to the multivariate case. *J. Multivariate Anal.*, 28(1):20–68.

Embrechts, P., Klüppelberg, C., and Mikosch, T. (1997). *Modelling Extremal Events*, volume 33 of *Applications of Mathematics (New York)*. Springer-Verlag, Berlin.

Erdélyi, A. (1956). *Asymptotic Expansions.* Dover Publications, Inc., New York.

Fan, J. and Yao, Q. (2003). *Nonlinear Time Series.* Springer Series in Statistics. Springer-Verlag, New York.

Feller, W. (1971). *An Introduction to Probability Theory and its Applications. Vol. II.* Second edition. John Wiley & Sons, Inc., New York-London-Sydney.

Freedman, D. and Lane, D. (1981). The empirical distribution of the Fourier coefficients of a sequence of independent, identically distributed long-tailed random variables. *Z. Wahrsch. Verw. Gebiete*, 58(1):21–39.

Georgiou, S. and Koukouvinos, C. (2006). Multi-level k-circulant supersaturated designs. *Metrika*, 64(2):209–220.

Gray, R. M. (2006). Toeplitz and Circulant Matrices: A review. *Foundations and Trends® in Communications and Information Theory*, 2(3):155–239.

Grenander, U. and Szegő, G. (1984). *Toeplitz Forms and their Applications*. Chelsea Publishing Co., New York, second edition.

Hoffman, A. J. and Wielandt, H. W. (1953). The variation of the spectrum of a normal matrix. *Duke Math. J.*, 20:37–39.

Kallenberg, O. (1986). *Random Measures*. Akademie-Verlag, Berlin; Academic Press, Inc., London, fourth edition.

Knight, K. (1991). On the empirical measure of the Fourier coefficients with infinite variance data. *Statist. Probab. Lett.*, 12(2):109–117.

LePage, R., Woodroofe, M., and Zinn, J. (1981). Convergence to a stable distribution via order statistics. *Ann. Probab.*, 9(4):624–632.

Lin, Z. and Liu, W. (2009). On maxima of periodograms of stationary processes. *Ann. Statist.*, 37(5B):2676–2695.

Massey, A., Miller, S. J., and Sinsheimer, J. (2007). Distribution of eigenvalues of real symmetric palindromic Toeplitz matrices and circulant matrices. *J. Theoret. Probab.*, 20(3):637–662.

Meckes, M. W. (2009). Some results on random circulant matrices. In *High Dimensional Probability V: the Luminy Volume*, volume 5 of *Inst. Math. Stat. (IMS) Collect.*, pages 213–223. Inst. Math. Statist., Beachwood, OH.

Mikosch, T., Resnick, S., and Samorodnitsky, G. (2000). The maximum of the periodogram for a heavy-tailed sequence. *Ann. Probab.*, 28(2):885–908.

Pollock, D. S. G. (2002). Circulant matrices and time-series analysis. *Internat. J. Math. Ed. Sci. Tech.*, 33(2):213–230.

Resnick, S. I. (1987). *Extreme Values, Regular Variation, and Point Processes*, volume 4 of *Applied Probability. A Series of the Applied Probability Trust*. Springer-Verlag, New York.

Riesz, M. (1923). Surle probleme des moments. Troisieme note (French). *Ark. Mat. Fys*, 16:1–52.

Samorodnitsky, G. and Taqqu, M. S. (1994). *Stable Non-Gaussian Random Processes*. Chapman & Hall, New York.

Sen, A. (2006). *Large Dimensional Random Matrices*. M. Stat Project Report. Indian Statistical Institute, Kolkata.

Soshnikov, A. (2004). Poisson statistics for the largest eigenvalues of Wigner random matrices with heavy tails. *Electron. Comm. Probab.*, 9:82–91.

Soshnikov, A. (2006). Poisson statistics for the largest eigenvalues in random matrix ensembles. In *Mathematical Physics of Quantum Mechanics*, volume 690 of *Lecture Notes in Phys.*, pages 351–364. Springer, Berlin.

Strok, V. V. (1992). Circulant matrices and spectra of de Bruijn graphs. *Ukrain. Mat. Zh.*, 44(11):1571–1579.

Walker, A. M. (1965). Some asymptotic results for the periodogram of a stationary time series. *J. Austral. Math. Soc.*, 5:107–128.

Weyl, H. (1916). Über die Gleichverteilung von Zahlen mod. Eins (German). *Math. Ann.*, 77(3):313–352.

Wu, Y. K., Jia, R. Z., and Li, Q. (2002). g-circulant solutions to the $(0, 1)$ matrix equation $A^m = J_n$. *Linear Algebra Appl.*, 345:195–224.

Zhou, J. T. (1996). A formula solution for the eigenvalues of g-circulant matrices (Chinese). *Math. Appl. (Wuhan)*, 9(1):53–57.

Index

$A_{k,n}$, additional properties, 83
$A_{k,n}$, dependent, 53, 62, 138
$A_{k,n}$, heavy tailed, 147, 155
$A_{k,n}$, point process, 130
$A_{k,n}$, scaled eigenvalues, 115
$A_{k,n}$, spectral radius, 89
C_n, LSD, 28
C_n, dependent, 39
J_n, 25
$Q_{h,4}$, 16
RC_n, LSD, 30
RC_n, dependent, 45
RC_n, heavy tailed, 144
RC_n, point process, 121
RC_n, scaled eigenvalues, 100
RC_n, spectral radius, 78
SC_n dependent, 48
SC_n heavy-tailed, spectral radius, 165
SC_n, LSD, 30
SC_n, extreme, 72
SC_n, heavy tailed, 145
SC_n, point process, 126
SC_n, scaled eigenvalues, 104
W_2-metric, 23
E_{F_n}, 11
$\Pi(w)$, 14, 15
$\Pi^*(w)$, 15
β_{2k}, 16
$\mathcal{W}_{2k}(2)$, 14, 16
$\mathcal{L}_R$, moment, 18
k-circulant, 3, 31, 50, 114
k-circulant, $n = k^2 + 1$, 115, 128
k-circulant, $n = k^2 + 1$, dependent, 137
k-circulant, $n = k^g + 1$, 117, 128, 145
k-circulant, $n = k^g - 1$, 154

k-circulant, $sn = k^g + 1$, 97
k-circulant, LSD, 34
k-circulant, additional properties, 83
k-circulant, eigenvalue proof, 173
k-circulant, radial component, 31
k-circulant, spectral radius, 88
w, number of distinct letters, 14
$w[i]$, 14
(M1), 16
(M1), circuit, 13
(M2), circuit, 14
(M4), 16, 17
(M4), circuit, 14

almost sure convergence, 178
alphabetical order, 14
asymptotics, Laplace, 79
autocovariance function, 37

Bernstein's inequality, 177
Berry-Esseen bound, 27, 179
bivariate normal, 39
Bonferroni inequality, 72, 123, 177
Boole's inequality, 177
Borel-Cantelli lemma, 11, 178
bound, Berry-Esseen, 27, 179

centering, 78
central limit theorem, 69, 178
central limit theorem, Lindeberg, 28
central limit theorem, stationary process, 182
characteristic polynomial, 5
Chebyshev's inequality, 56, 177
circle, 31
circuit, 13
circuit, $\Pi(w)$, 14
circuit, (M1), 13

circuit, (M2), 14
circuit, (M4), 14
circuit, equivalence, 14
circuit, matched, 14
circuit, multiple, 14
circuit, pair-matched, 14
circulant, 1, 28, 38, 67, 100, 160
condition, minimal, 24
convergence in distribution, 178
convergence in probability, 178
convergence of types, 92, 180
convergence, almost sure, 178
cross-matched, 14

dependent $A_{k,n}$, 138
dependent input, heavy tailed, 166
distribution, convergence, 178
distribution, exponential, 45
distribution, Gumbel, 31, 45, 67, 79,
        90, 180
distribution, heavy-tailed , 181
distribution, stable, 139, 181
distribution, symmetrized Rayleigh,
        18
domain of attraction, 139, 140
dominated convergence theorem, 57,
        178
doubly symmetric Hankel matrix, 23
doubly symmetric Toeplitz matrix,
        22

eigenvalue, scaled, 99
empirical spectral distribution, 9
equivalence class, 14
equivalence relation, 13
equivalent circuits, 14
Euclid, 51
exceedance, 125
expected spectral distribution
        function, 9
exponential distribution, 45
exponential variable, 121, 125, 128
exponentials, product, 79, 83, 90
extreme order statistics, 79
extreme value, 67

extreme, SC_n, 72

Fourier frequency, 37

generating vertex, 15
Gumbel distribution, 31, 45, 67, 90,
        180
Gumbel limit, periodogram, 182
Gumbel, max-domain, 79

Hankel, doubly symmetric, 22
heavy-tailed distribution, 181
heavy-tailed input, 160
heavy-tailed, dependent input, 166
Hoffmann-Wielandt inequality, 24
Holder's inequality, 16, 177

inequality, Bernstein, 177
inequality, Bonferroni, 72, 123, 177
inequality, Boole, 177
inequality, Chebyshev, 56, 177
inequality, Hoffmann-Wielandt, 24
inequality, Holder, 16, 177
inequality, interlacing, 26
inhomogeneous Poisson, 120
input sequence, 12
intensity function, 121, 126, 130, 135,
        137, 138
interlacing inequality, 26

jointly-matched, 14

Karamata's theorem, 162

Laplace's asymptotics, 79
light tail, 99
light-tailed LSD, 156
limiting moment, 16
limiting spectral distribution, 10
Lindeberg's central limit theorem, 28
linear process, 37
link function, 13
LSD, C_n, 28
LSD, k-circulant, 34
LSD, RC_n, 30
LSD, SC_n, 30

LSD, dependent $A_{k,n}$, 53, 62
LSD, dependent C_n, 39
LSD, dependent RC_n, 45
LSD, dependent SC_n, 48
LSD, light-tailed, 156
LSD, mixture, 38

major partition, 146
matched circuit, 14
matched, crossed, 14
matched, jointly, 14
matched, location, 14
matched, self, 14
matrix, doubly symmetric Hankel, 23
matrix, doubly symmetric Toeplitz, 22
matrix, palindrome, 23
max-domain, 79
max-domain, product of exponentials, 83
maxima, normal random variable, 70
maximum, periodogram, 160
metric, W_2, 23
metrizable, 23
Mill's ratio, 106
minima, normal random variable, 70
minimal condition, 24
moment generating function, 70
moment method, 10
moment, $\mathcal{L}_R$, 18
moment, limiting, 16
moment, reduced, 23
moving average, two-sided, 37, 99, 100, 104, 115, 117
multiple circuits, 14

non-generating vertex, 15
normal approximation, 27, 87, 112
normal distribution, max domain, 181
normal moments, Riesz's condition, 11
normal random variable, maxima, 70
normal random variable, minima, 70
norming constants, 181

ordered eigenvalues, 125

pair-matched circuit, 14
pair-matched word, 14, 15
palindrome, 23
partial sum, 41
partition, 155
partition block, 14
partition, major, 146, 154
partition, singleton, 145
periodogram, 37, 160
periodogram, Gumbel limit, 182
periodogram, heavy-tailed, 160
point measure, 119, 181
point process, $A_{k,n}$, 130
point process, RC_n, 121
point process, SC_n, 126
point process, Poisson, 182
point process, simple, 119, 182
Poisson point process, 121, 126, 130, 133, 135, 137, 138
polar coordinates, 54
probability inequalities, 177
process, stationary, 179
product of exponentials, 79, 83, 90

Rayleigh, symmetrized, 30
reduced moment, 23
relatively prime, 84
reverse circulant, 3, 17, 67, 100, 104, 160
reverse circulant limit, $\mathcal{L}_R$, 18
reverse circulant, heavy tail, 144
reverse circulant, normal approximation, 17
reverse circulant, point process, 120
Riesz's condition, 11, 16
Riesz's condition, normal moments, 11
root of unity, 2

scaled eigenvalue, 99
scaled eigenvalues, $A_{k,n}$, 115
scaled eigenvalues, RC_n, 100
scaled eigenvalues, SC_n, 104
scaling, 12, 78, 159

self-matched, 14
simple point process, 119, 182
simulation, $\frac{1}{\sqrt{n}}A_{k,n}$, 31, 65
simulation, $\frac{1}{\sqrt{n}}C_n$, 29, 40
simulation, $\frac{1}{\sqrt{n}}RC_n$, 17, 46
simulation, $\frac{1}{\sqrt{n}}SC_n$, 20, 48
spectral density, 37, 179
spectral distribution, 9, 179
spectral norm, 78
spectral radius, 67
spectral radius, $A_{k,n}$, 89
spectral radius, k-circulant, 88
spectral radius, RC_n, 78
spectral radius, SC_n heavy-tailed,
        165
stable distribution, 139, 181
stable, domain of attraction, 139, 140
standard normal, 38
stationary process, 179
stationary process, central limit
        theorem, 182
stationary, weakly, 179
strong law of large numbers, 179
symmetric circulant, 2, 20, 47, 104,
        105, 126, 171
symmetric circulant, dependent, 137
symmetric circulant, heavy tail, 144
symmetric word, 17
symmetrized Rayleigh, 18, 30

tail, 67
tail of product, 79
tail, light, 99, 156
Toeplitz, doubly symmetric, 22
topology, vague, 119
trace-moment formula, 11, 13
truncation, 86, 111
two-sided moving average, 37, 99,
        100, 104, 115, 117

unit circle, 34, 62

vague convergence, 119, 181
vague topology, 119
vertex, 14

vertex, generating, 15
vertex, non-generating, 15
Von Mises function, 180

weak convergence, almost surely, 10
weak convergence, in probability, 10
weakly stationary, 179
Weyl's result, 164
white noise, 180
word, pair-matched, 14, 15
word, symmetric, 17